The Genera of Australian Lichens

The Genera of Australian Lichens

(LICHENIZED FUNGI)

Roderick W. Rogers

University of Queensland Press
St Lucia • London • New York

Typeset by Press Etching Pty Ltd, Brisbane
Printed and bound by Silex Enterprise & Printing Co., Hong Kong

Distributed in the United Kingdom, Europe, the Middle East, Africa, and the Caribbean by Prentice-Hall International, International Book Distributors Ltd, 66 Wood Lane End, Hemel Hempstead, Herts., England.

National Library of Australia
Cataloguing-in-Publication data

Rogers, Roderick W. (Roderick Westgarth), 1944-
The genera of Australian lichens (lichenized fungi).

ISBN 0 7022 1579 1

1. Lichens – Australia – Classification.
I. Title.

589.1'012

TO NAOMI
who watched
and asked
and encouraged

Contents

Figures

Acknowledgments

The preparation of this work has been assisted by many people, and I am grateful to them all. Special mention should, however, be made of some. The late Dr G.C. Bratt, Dr J.A. Elix and Mr N. Sammy all contributed information concerning genera known from their regions but not previously reported in the literature. Mr R.B. Filson of the National Herbarium, Melbourne, contributed a number of suggestions at various stages of the preparation of the manuscript which materially improved it. The keys were tested and refined by the small community of lichenologists in Australia, including not only those mentioned above but also Mr D. Archer and Ms N. Stevens. Mr Colin Beasley made an important contribution in the very early stages of the preparation of the work collecting and translating genus descriptions from a variety of sources.

This account could not have been prepared at all had not the Interim Council of the Australian Biological Resources Study funded a survey of the lichens in the Brisbane area which necessitated gathering most of the information contained in this volume.

Introduction

This account of the genera of lichen forming fungi in Australia has been prepared so that the naturalist anywhere in Australia can name a lichen to genus. References have been provided to the scattered literature which will allow species determinations in many genera.

There is no extensive account of lichen biology in this book. Those who wish to study lichens are referred to the introductory accounts in Filson and Rogers' *Lichens of South Australia* or to the fuller accounts in Hale's *The Biology of Lichens*. The reading of such an account will greatly facilitate use of this set of keys and descriptions. The glossary is full, refers to a number of illustrations, and certainly should be sufficient for those with some knowledge of the fungi.

The Nature of Lichens

A lichen is probably best regarded not as an organism but as a tightly integrated biological community. A lichen always consists of at least one species of alga (represented by numerous individuals) and one species of fungus (possibly representing more than one individual). There are a number of lichens which are reported to characteristically posses two algal species either as alternatives or simultaneously, especially as cephalodia.

Such a complex community cannot be given a species name under the International Code of Botanical Nomenclature, but perhaps it could within a phytosociological framework.

The International Code of Botanical Nomenclature is quite clear that the name given to a lichen applies only to its fungal component: "For nomenclatural purposes the names given to Lichens shall be considered as applying to their fungal symbionts" (Article 13d).

The state of knowledge of lichens is not, however, so well advanced as to allow a full integration of the lichen classification with that of the other fungi. With some exceptions lichenization is taken as a character to separating fungal orders.

Sources of Information

The genera included in this account are those listed for the region by Weber and Wetmore (1972), Wetmore (1963) or reported in the literature or communicated personally by various workers. Those genera in the latter categories are listed elsewhere.

A convenient starting point in the accumulation of descriptions of the genera was the account prepared by Zahlbruckner (1903). That account provided a preliminary description for almost all of the genera. A second major source was Santesson (1952) which summarized all the foliicolous genera, and commented on related genera in a most helpful manner. Recourse has been taken also to a wide range of check lists and regional floras to ensure that genera have been used in a currently acceptable sense. Not all genera, however, are currently treated in a like manner by all writers. A particularly difficult group is the genus *Parmelia sensu lato*, which has been split into a large number of segregates each with genus status by Mason Hale (1974a,b,c) and Theodore Esslinger (1978a). In this arrangement I have followed Hale and Esslinger, not because I am convinced that all the segregates are good genera, but because at least some are. This arrangement makes a maximum of information available.

In recent times there has been a great deal of work done on the delimitation of the genera of crustose lichens. In particular the large genera *Lecanora* and *Lecidea* together with the other superficially similar genera, have been reviewed. As a result many old genera have been reinstated for use, and many new genera proposed. Our knowledge of the crustose lichens of Australia is very scant indeed, and it is not possible to be sure which of these many genera do occur in Australia. A conservative approach to the taxonomy of the crustose lichens has therefore been taken: perhaps at some time in the future a critical review like that of the British lichen flora by Hawksworth, James and Coppins (1980) will be possible in Australia.

Extra information for most genera has been drawn from the account of families prepared by Poelt (1973), providing information which will not normally be used in identifying material (for example, nature of ascus wall) but which does help make the arrangement into families more clear.

It is doubtful if very much significance can be attached to the description of fulcrum form in pycnidia, but the form of pycnospore has been used to separate genera, notably in division of the old genus *Physcia*. The information has therefore been provided for all genera possible.

Other pieces of information have been gleaned from diverse literature sources over a number of years — in some cases without record of the sources being retained. Original literature sources are therefore not normally acknowledged.

Information on the chemistry of genera has been compiled in the first instance from the accounts by Culberson (1969, 1970) and Culberson *et al.* (1977). In a number of cases, however, revisional studies have either superseded or supplemented their information.

Species number in each genus is based usually on the figure given by Poelt (1973), but figures from revisional studies have sometimes been used instead. Australian distributions are based on the information in Weber and Wetmore (1972) and Wetmore (1963), supplemented by revisional studies, personal communications and the author's field knowledge.

Many pieces of information have been gleaned from diverse sources over a period of years, and a debt is owned to a great many lichenologists whom it is no longer possible to acknowledge directly.

Notes on the Use of Keys

Users are reminded that there are seven separate keys. Some genera occur in more than one key, for the criteria separating the form groups are not clear cut. Some genera will be found treated in sections to which they strictly do not belong, but to which a casual examination will suggest they do. The keys are designed not to reflect taxonomy, but simply to facilitate naming.

The keys to the fruticose and foliose genera have been designed to allow determination of a specimen when sterile as far as is practicable — and many species will be found only in the sterile condition. The keys to other groups could not be so arranged. It is simply not possible to key out a sterile crustose or squamulose lichen at this time. We do not have sufficient information, and it seems unlikely that we ever will.

To key out to genus a crustose or squamulose lichen requires fertile material, and access to a microscope. Such determinations can only be made in a laboratory, not in the field.

The wording of the keys is deliberate. Users must read both of the choices offered in these strictly dichotomous keys before deciding which is the appropriate one, and must consider all of the information included. Until the user is familiar with the usage of words in the keys, free use should be made of the glossary. When a genus name has been arrived at from the key, the specimen should be carefully compared with the description, for a key cannot always provide the right answer even if the user has made the correct choice at every step — and since incorrect choices are possible, to do so is even more important. Further, comparison with reliably named material in a herbarium is most desirable.

Genera New for Australia

The following genera were not listed for Australia in the catalogues produced by Wetmore (1963) for Tasmania and Weber and Wetmore (1972) for Australia excluding Tasmania, but are now known for the region.

Those marked * are the result of entirely new records, and the others are genera segregated out of those indicated.

***Arctomia*
Bulbothrix (Parmelia)
**Candelariella*
**Cetrelia*
Chrysothrix (Lepraria)
Cladina (Cladonia)
Coccotrema (Perforaria)
**Conotremopsis*
**Cystocoleus*
Diploicia (Buellia)
Dirinaria (Physcia)
Fistulariella (Ramalina)
Heterodermia (Anaptychia)
Hypotrachyna (Parmelia)
Mallotium (Leptogium)
**Mastodia*
Melanelia (Parmelia)
Neofuscelia (Parmelia)
Pannoparmelia (Anzia)
Parmelina (Parmelia)
Parmotrema (Parmelia)
Phaeophyscia (Physcia)
Physciopsis (Physcia)
**Placynthium*
**Polycaulonia*
Protoblastenia (Lecidea)
Pseudoparmelia (Parmelia)
Psora (Lecidea)
**Psoromaria*
Relicina (Parmelia)
**Sagenidium*
**Spilonema*
Squamarina (Lecanora)
**Thelidium*
**Thyrea*
**Trimmatothele*
Xanthoparmelia (Parmelia)
**Xylographa*

Classification

This outline classification of the genera of lichen forming fungi reported from Australia is based on Poelt (1973).

DIVISION EUMYCOTA
SUBDIVISION ASCOMYCOTINA

ORDER ARTHONIALES
ARTHONIACEAE
Arthonia
Arthothelium
Gymnographa
OPEGRAPHACEAE
Chiodecton
Enterodictyon
Enterographa
Mazosia
Opegrapha
LECANACTIDACEAE
Catinaria
Lecanactis
Melampydium
Schismatomma
ROCELLACEAE
Dirina
Dirinastrum
Rocella
Sagenidium
CHRYSOTHRICACEAE
Chrysothrix

ORDER DOTHIDALES
PLEOSPORACEAE
Arthopyrenia
Dermatina
Microthelia
Mycoporellum
Polyblastiopsis
Tomasellia

ORDER VERRUCARIALES
VERRUCARIACEAE
Dermatocarpon
Endocarpon
Mastodia
Thelidium
Trimmatothele
Verrucaria

ORDER PYRENULALES
PYRENULACEAE
Anthracothecium
Asteroporum
Parathelium
Parmentaria
Pleurotheliopsis
Pseudopyrenula
Pyrenula
TRYPETHELIACEAE
Melanotheca
Trypethelium
LAURERACEAE
Bottaria
Campylothecium
Laurera
STRIGULACEAE
Aspidothelium
Pleurotrema
Strigula
CLATHROPORINACEAE
Clathroporina
Porina
Tricothelium

ORDER CALICIALES
MYCOCALICIACEAE
Stenocybe
CALICIACEAE
Calicium
Chaenotheca
Coniocybe
Cyphelium
Pyrgillus
Tylophoron
SPHAEROPHORACEAE
Sphaerophorus

ORDER OSTROPALES
THELOTREMACEAE
Conotremopsis
Diploschistes
Leptotrema
Ocellularia
Phaeotrema
Thelotrema
Tremotylium

ORDER GRAPHIDALES
GRAPHIDACEAE
Aulaxina
Diplogramma
Glyphis
Graphina
Graphis
Helminthocarpon
Medusulina
Phaeographina
Phaeographis
Sarcographa
Sarcographina
MELASPILEACEAE
Melaspilea
ASTROTHYRIACEAE
Echinoplaca
Gyalectidium
Sagiolechia
Tricharia
GYALECTACEAE
Coenogonium
Dimerella
Gyalecta

ORDER LECANORALES
LICHINACEAE
Ephebe
Lempholemma
Lichina
Peccania
Porocyphus
Pyrenopsidium
Pyrenopsis
Synalissa
Thyrea
HEPPIACEAE
Heppia
Peltula
PLACYNTHIACEAE
Placynthium
Psoroma
Psoromaria
PELTIGERACEAE
Peltigera
NEPHROMIACEAE
Nephroma
LOBARIACEAE
Lobaria
Pseudocyphellaria
Sticta
COLLEMATACEAE
Collema
Leptogium
Mallotium
Physma
COCCOCARPIACEAE
Coccocarpia
Spilonema
PANNARIACEAE
Erioderma
Pannaria
Parmeliella
ARCTOMIACEAE
Arctomia
LECIDEACEAE
Bacidia
Bombyliospora
Byssoloma
Catillaria
Crocynia
Lecidea
Lopadium
Megalospora
Phylctella
Phyllopsora
Psora
Rhizocarpon
Sporopodium
Tapellaria
Toninia

LECANORACEAE
Haematomma
Lecania
Lecanora
Squamarina
ASPICILIACEAE
Aspicilia
Placopsis
HYPOGYMNIACEAE
Hypogymnia
Menegazzia
Pseudevernia
PARMELIACEAE
Bulbothrix
Cetraria
Cetrelia
Chondropsis
Hypotrachyna
Pannoparmelia
Parmelia
Parmelina
Parmeliopsis
Parmotrema
Pseudoparmelia
Relicina
Xanthoparmelia
USNEACEAE
Alectoria
Coelocaulon
Evernia
Neuropogon
Usnea
RAMALINACEAE
Fistulariella
Ramalina
ANZIACEAE
Anzia
STEREOCAULACEAE
Pilophorus
Stereocaulon
CLADONIACEAE
Cladia
Cladina
Cladonia
Gymnoderma
Ramalea
Thysanothecium
HETERODEACEAE
Heterodea
BAEOMYCETACEAE
Baeomyces
SIPHULACEAE
Siphula
Thamnolia
UMBILICARIACEAE
Umbilicaria
ACAROSPORACEAE
Acarospora
Biatorella
Maronea
Sarcogyne
Sporastatia
PERTUSARIACEAE
Coccotrema
Ochrolechia
Pertusaria
CANDELARIACEAE
Candelaria
Candelariella
TELOSCHISTACEAE
Caloplaca
Fulgensia
Polycaulonia
Protoblastenia
Teloschistes
Xanthoria
PHYSCIACEAE
Anaptychia
Buellia
Diploicia
Dirinaria
Heterodermia
Phaeophyscia
Physcia
Physciopsis
Pyxine
Rinodina
MICAREACEAE
Micarea
TRAPELIACEAE
Placopsis
Trapelia
AGYRACEAE
Xylographa

SUBDIVISION BASIDIOMYCOTINA
ORDER APHYLLOPHORALES
DICTYONEMATACEAE
Dictyonema
Herpothallon

SUBDIVISION DEUTEROMYCOTINA
Cystocoleus
Dendriscocaulon
Leparia
Leprocaulon
Normandina

Keys to the Genera of Lichens in Australia

A series of seven artificial keys to genera are presented:

In some cases it has not been possible to provide key characters for genera. These are a few genera of crustose or squamulose lichens which are not known to form fruiting bodies, the basis for the key to those groups. These genera are listed in the end of the relevant key.

Fruticose thalli

1.	Thallus filamentous or terete, regular	2
1.	Thallus flattened or irregular and of unequal thickness	9
2.	With a resistant chondroid axis	3
2.	Without a chondroid axis	4
3.	Branches light yellow at the base, shading to black in the upper regions, disk black or dark	*Neuropogon*
3.	Branches yellow-green to grey or red, disk pale	*Usnea*
4.	Thallus hollow, or with perforations through the cortex to the central medullary regions	43
4.	Thallus solid, imperforate	5
5.	Thallus yellow-green to grey or pink	45
5.	Thallus brown or black (at least near the tips)	6
6.	Without pseudocyphellae, thallus black (at least at the tips)	7
6.	With pseudocyphellae, thallus brown	47
7.	Thallus of blue-green algal filaments (*Stigonema*) with hyphae ramifying through the gelatinous sheath	8
7.	Thallus with a well developed fungal cortex enclosing the green alga *Trebouxia* or the blue-green alga *Nostoc*	48
8.	Apothecia immersed in swellings	*Ephebe*
8.	Apothecia terminal on lobes	*Spilonema*
9.	Thallus gelatinous when wet	10
9.	Thallus not gelatinous when wet	14
10.	Thallus a loose weft of prostrate filaments	11
10.	Thallus not of prostrate filaments	12
11.	Apothecia immersed in swellings	*Ephebe*
11.	Apothecia terminal on lobes	*Spilonema*
12.	On rocks in the marine splash zone, apothecia in urceolate swellings	*Lichina*
12.	Not in marine splash zone, apothecia not in urceolate swellings	13
13.	Thallus of thin ascending branches (usually on granites)	*Porocyphus*
13.	Thallus branches as thick as they are tall, aggregated to form a crust-like plaque (usually on calcareous substrates)	*Synalissa*

14. Crustose primary thallus bearing minutely stipitate fruiting bodies (less than 2 mm tall) — 15
14. Thallus entirely fruticose, or the primary thallus bearing podetia or pseudopodetia more than 2 mm tall — 18

15. Spores simple — 16
15. Spores septate — 17

16. Spores pale (hyaline to yellow) — *Coniocybe*
16. Spores dark (brown) — *Chaenotheca*

17. Spores once septate — *Calicium*
17. Spores 3-7 septate — *Stenocybe*

18. Thallus hollow — 19
18. Thallus solid — 23

19. Thallus vermiform, white or pale grey, alpine — *Thamnolia*
19. Thallus not vermiform, variously coloured and distributed — 20

20. On twigs, bark or wood — *Fistulariella*
20. On soil or rocks, or among mosses and litter — 21

21. Medulla stained yellow with usnic acid, on acid heath soils — *Ramalea*
21. Medulla not yellow from usnic acid — 22

22. Walls perforate to the hollow centre or rather lax medulla — *Cladia*
22. Axils sometimes perforate, but not the lateral walls — *Cladonia*

23. Spores free in a mazaedium at maturity — *Sphaerophorus*
23. Spores retained in an ascus at maturity — 24

24. Branches with a strong central chondroid axis — 25
24. Branches without a strong central chondroid axis — 26

25. Branches light yellow at the base becoming black at the tips — *Neuropogon*
25. Branches yellow-green to grey or red — *Usnea*

26. Apothecia lateral on the expanded top of the flattened podetia — *Thysanothecium*
26. Apothecia not on expanded, flattened podetia — 27

27. Thallus of well developed basal squamules bearing marginal podetia — 28
27 Thallus not of squamules bearing marginal podetia, but cylindrical, irregular or strap-like — 29

28. On acid heathland soils, podetia a few millimetres tall — *Ramalea*
28. On trees in wet forests, podetia to 10 millimetres tall — *Gymnoderma*

29. Thallus more or less cylindrical or irregular in cross-section — 30
29. Thallus bilaterally flattened, strap-like — 34

30. Thallus of anastomosing prostrate pseudocyphellae, rather rope-like strands on calcareous arid soils — *Aspicilia*
30. Thallus erect, not on arid soils — 31

31. Lobes with phyllocladia — *Stereocaulon*
31. Lobes without phyllocladia, but may be sorediate — 32

32. Thallus golden at least in parts, without a crustose primary thallus — 33
32. Thallus never golden, with a primary crustose thallus — 40

33. Thallus of recumbent lobes usually 1 cm or more long with lateral apothecia — *Teloschistes*
33. Thallus of short (few millimetres) long lobes with terminal apothecia — *Polycaulonia*

34. Thallus gold, apothecia K+ purple (with parietin) — 35
34. Thallus not gold, apothecia not K+ purple (no parietin) — 36

35. Thallus of recumbent lobes usually 1 cm or more long with lateral apothecia — *Teloschistes*
35. Thallus of short (few millimetres) long lobes with terminal apothecia — *Polycaulonia*

36. On mangroves on tropical coasts — 37
36. Not on tropical mangroves — 38

37. Thallus white or grey — *Rocella*
37. Thallus straw-yellow to green — *Ramalina*

38. Thallus straw-yellow to green — *Ramalina*
38. Thallus white to silver grey or brown — 39

39. Thallus esorediate — 40
39. Thallus sorediate — *Evernia*

40. Apothecia immersed in terminal swellings — *Sphaerophorus*
40. Apothecia not immersed — 41

41. On maritime rocks, apothecia adnate — *Sagenidium*
41. On alpine rocks, peat or on soil or twigs — 42

42. Thallus less than 2 cm tall, base of lobes immersed in soil or peat — 46
42. Thallus larger, loosely adnate on rocks, soil or twigs — *Cetraria*

43. Thallus perforate — *Cladia*
43. Thallus imperforate — 44

44. Apothecia naked, in terminal cymes, apices nodding — *Cladina*
44. Apothecia immersed in terminal swellings, apices erect — *Sphaerophorus*

45. Apothecia sessile or shortly stipitate, with a distinct margin — *Ramalina*
45. Apothecia on erect podetia, often inflated and emarginate — *Baeomyces*

46. Thallus brown, usually with black apothecia — *Toninia*
46. Thallus white or grey, always sterile — *Siphula*

47. Phycobiont green, thallus little branched — *Coelocaulon*
47. Phycobiont blue-green, thallus much branched — *Dendriscocaulon*

48. Phycobiont green, thallus cylindrical or terete — *Alectoria*
48. Phycobiont blue-green, thallus irregular — *Dendriscocaulon*

Foliose thalli

1.	Lower surface with well developed rhizines	2
1.	Lower surface apparently devoid of rhizines, sparsely rhizinate at the margins only, naked or tomentose	44
2.	Lobes 1 cm or more wide	3
2.	Lobes less than 1 cm wide	15
3.	Lower cortex with cyphellae or pseudocyphellae	4
3.	Lower cortex with neither cyphellae nor pseudocyphellae	5
4.	Lower cortex truly cyphellate	*Sticta*
4.	Lower cortex pseudocyphellate	*Pseudocyphellaria*
5.	Lower surface with vein-like markings, apothecia on the upper side of marginal projections	*Peltigera*
5.	Not as above	6
6.	Upper cortex yellow-gold, K+ purple (parietin)	*Xanthoria*
6.	Upper cortex brown, grey or straw-yellow to green, never K+ purple (no parietin)	7
7.	Thallus brown	70
7.	Thallus grey or straw-yellow to green, apothecia on the upper surface	8
8.	Upper cortex pseudocyphellate or reticulately cracked on young lobes	9
8.	Upper cortex not pseudocyphellate nor reticulately cracked except on very old lobes	10
9.	Pseudocyphellae minute (less than 0.1 mm diameter) ovoid	*Cetrelia*
9.	Pseudocyphellae larger, effigurate	*Parmelia*
10.	On rocks or soil, thallus straw-yellow to green (usnic acid)	*Xanthoparmelia*
10.	On any substrate, but if on rocks or soil not straw-yellow to yellow-green (without usnic acid)	11
11.	Rhizines dichotomously branched	*Hypotrachyna*
11.	Rhizines simple or squarrosely branched	12
12.	Thallus with marginal cilia	13
12.	Thallus without marginal cilia	14
13.	Lobes apically truncate, rhizines to the lobe tips	*Parmelina*
13.	Lobes broadly rotund at the tip, rhizines absent from a broad zone below the tips of lobes	*Parmotrema*

14.	Lobes apically truncate, rhizines present to the lobe tips	*Pseudoparmelia*
14.	Lobes broadly rotund at the tips, rhizines absent from a broad zone below the tips of the lobes	*Parmotrema*
15.	Margin of the thallus ciliate	16
15.	Margin of the thallus eciliate	25
16.	Cilia bulbate at the base	17
16.	Cilia not bulbate	18
17.	Thallus grey (atranorin)	*Bulbothrix*
17.	Thallus straw-yellow to yellow-green (usnic acid)	*Relicina*
18.	Upper cortex brown	*Phaeophyscia*
18.	Upper cortex white, grey or straw yellow to green	19
19.	Upper cortex of young lobes reticulately cracked or pseudocyphellate	*Parmelia*
19.	Upper cortex of young lobes neither reticulately cracked nor pseudocyphellate	20
20.	Rhizines dichotomously branched	21
20.	Rhizines simple or squarrose	23
21.	Upper cortex of longitudinally oriented hyphae, giving the thallus a waxy appearance	22
21.	Upper cortex of plectenchyma, giving the thallus a dull surface	*Hypotrachyna*
22.	Spores with blastidia, thallus normally adnate	*Heterodermia*
22.	Spores with two locules only, thallus normally ascending	*Anaptychia*
23.	Spores two celled, lobes usually 2 mm or less wide	*Physcia*
23.	Spores simple, lobes usually more than 2 mm wide	24
24.	Lobe tips truncate, rhizines right to the tips of lobes	*Parmelina*
24.	Lobe tips broadly rotund, rhizines absent from a broad marginal zone	*Parmotrema*
25.	Upper surface bright yellow or yellow gold	26
25.	Upper surface, white, grey, brown or straw-yellow to yellow-green	27
26.	Thallus K+ purple (parietin), lobes more than 2 mm wide	*Xanthoria*
26.	Thallus K− (without parietin), lobes less than 1 mm broad	*Candelaria*
27.	Thallus brown or olive	28
27.	Thallus white, grey or straw-yellow to yellow-green	31
28.	Spores simple	29
28.	Spores septate	30
29.	Upper cortex of interwoven hyphae	*Parmeliopsis*
29.	Upper cortex plectenchymatous	74
30.	Pycnospores spherical (lobes usually about 2 mm broad)	*Phaeophyscia*
30.	Pycnospores acicular (lobes often less than 1 mm broad)	*Physciopsis*
31.	Thallus on rocks or soil and straw-yellow to yellow-green	*Xanthoparmelia*
31.	Thallus white or grey, or if straw-yellow to yellow-green not on rocks or soil	32

32. Upper cortex of young lobes reticulately cracked or pseudocyphellate	33
32. Upper cortex of young lobes neither reticulately cracked nor pseudocyphellate	35
33. Lobes more than 2 mm wide	34
33. Lobes less than 2 mm wide	*Pyxine*
34. Pseudocyphellae minute (less than 0.1 mm diameter) ovoid	*Cetrelia*
34. Pseudocyphellae effigurate	*Parmelia*
35. Rhizines dichotomously branched	*Hypotrachyna*
35. Rhizines simple or squarrose	36
36. Upper cortex of interwoven hyphae	37
36. Upper cortex plectenchymatous	39
37. Spores simple	*Parmeliopsis*
37. Spores septate	38
38. Spores with blastidia, thallus normally adnate	*Heterodermia*
38. Spores without blastidia, thallus commonly ascending	*Anaptychia*
39. Spores simple	40
39. Spores septate	41
40. Lobes broadly rotund at the tips, rhizines absent from a broad naked zone near the lobe tips	*Parmotrema*
40. Lobes truncate at the tips, rhizines present right up to the tips of lobes	*Pseudoparmelia*
41. Apothecium with a proper exciple only	*Pyxine*
41. Apothecium with a thalloid exciple	42
42. Hypothecium hyaline	*Physcia*
42. Hypothecium dark	43
43. Thallus K+ yellow (atranorin)	*Dirinaria*
43. Thallus K− (no atranorin)	*Pyxine*
44. Rhizines present but restricted to the thallus margin	45
44. Rhizines absent	46
45. Spores with blastidia, thallus normally adnate	*Heterodermia*
45. Spores two celled only, thallus normally ascending	*Anaptychia*
46. Thallus umbilicate	47
46. Thallus not umbilicate	49
47. Thallus not gelatinous when wet	*Umbilicaria*
47. Thallus gelatinous when wet	48
48. Phycobiont *Scytonema*	*Porocyphus*
48. Phycobiont *Xanthocapsa*	*Thyrea*
49. Thallus on a tomentum	50
49. Thallus naked below	54
50. Upper surface of the thallus scrobiculate	*Lobaria*
50. Upper surface smooth or slightly irregular	51

51. Thallus with green algae	52
51. Thallus with blue-green algae	53
52. Lobes less than 2 mm broad	71
52. Lobes more than 2 mm broad	*Heterodea*
53. Phycobiont *Scytonema*	76
53. Phycobiont *Nostoc*	72
54. Phycobiont green	55
54. Phycobiont blue-green	62
55. Thallus inflated, hollow, or the medulla thick, often loose and cobweb-like	56
55. Thallus not inflated, solid, the medulla thin	57
56. Upper cortex pierced by holes to the centre of the thallus	*Menegazzia*
56. Upper cortex not pierced by holes	*Hypogymnia*
57. Thallus free or loosely attached to the soil	58
57. Thallus adnate on rock wood or bark	59
58. With cyphella-like pores below, apothecia marginal	*Heterodea*
58. Lower surface without cyphella-like pores, apothecia laminal	73
59. Upper cortex densely tomentose	*Erioderma*
59. Upper cortex naked	60
60. Thallus brown or olive	67
60. Thallus grey	61
61. Thallus closely adnate, medulla thin	*Dirinaria*
61. Thallus loosely adnate, medulla thick	69
62. Spores with a thick wall and slime coating	*Physma*
62. Spores with thin walls without slime	63
63. Thallus of hyphae wholly embedded in the algae	*Lempholemma*
63. Thallus with a developed cortex	64
64. Spores non-septate, thallus heteromerous	65
64. Spores septate, thallus homeomerous	66
65. Apothecia with a proper exciple	*Parmeliella*
65. Apothecia with a thalloid exciple	*Pannaria*
66. Upper cortex of interwoven hyphae	*Collema*
66. Upper cortex cellular	*Leptogium*
67. Thallus ascending, attached only by a holdfast	*Cetraria*
67. Thallus closely adnate, attached by the lower surface	68
68. Thallus with cephalodia	*Placopsis*
68. Thallus without cephalodia	*Physciopsis*
69. Lower surface black, thallus inflated	*Hypogymnia*
69. Lower surface pale, thallus compressed	*Cetraria*
70. Apothecia on upper surface of thallus	74
70. Apothecia on lower surface of thallus	*Nephroma*

71. Thallus white or grey, spores many per ascus *Anzia*
71. Thallus brown or green, spores eight per ascus *Pannoparmelia*

72. Upper cortex a single cell layer, thallus gelatinous, tomentum short *Mallotium*
72. Upper cortex several cell layers, thallus not gelatinous, tomentum often long 75

73. Thallus rolling into a ball when dry, pale below *Chondropsis*
73. Thallus often markedly channelled, but not rolling into a ball, dark below *Pseudevernia*

74. Thallus blue with HNO_3, with neither isidia nor soredia *Neofuscelia*
74. Thallus HNO_3- or HNO_3 red, usually with isidia or sorediate isidia *Melanelia*

75. Spores thin walled *Pannaria*
75. Spores thick walled *Physma*

76. Upper cortex smooth, striate *Coccocarpia*
76. Upper cortex shaggy *Erioderma*

Squamulose thalli

1.	Fruiting bodies markedly stipitate (with a secondary thallus)	2
1.	Fruiting bodies sessile to immersed	8
2.	Secondary thallus hollow	3
2.	Secondary thallus solid	4
3.	Hypothecium pale, apothecium flat or convex	*Cladonia*
3.	Hypothecium dark, apothecium conical or globose	*Pilophorus*
4.	Apothecia lateral on the tips of flattened podetia	*Thysanothecium*
4.	Apothecia terminal on more or less cylindrical podetia	5
5.	Medulla with usnic acid	*Ramalea*
5.	Medulla without usnic acid	6
6.	Spores septate	27
6.	Spores simple	7
7.	Primary thallus corticate, paraphyses reticulate	*Gymnoderma*
7.	Primary thallus ecorticate, paraphyses unbranched	*Pilophorus*
8.	Asci immersed in fruiting bodies opening by a pore	9
8.	Asci in open disk	11
9.	Spores simple	10
9.	Spores septate	*Endocarpon*
10.	Thallus ecorticate, of loosely woven hyphae	*Mastodia*
10.	Thallus corticate above, paraplectenchymatous	*Dermatocarpon*
11.	Spores simple	12
11.	Spores septate	24
12.	Asci polysporous	13
12.	Asci with 1-16 spores	15
13.	Thallus with a thick upper cortex	14
13.	Thallus ecorticate	*Pyrenopsis*
14.	Phycobiont blue-green	*Peltula*
14.	Phycobiont green	*Acarospora*
15.	Thallus ecorticate	*Pyrenopsis*
15.	Thallus with a developed cortex	16
16.	Apothecia immersed	17
16.	Apothecia sessile or adnate	18

17. Phycobiont blue-green	*Heppia*
17. Phycobiont green	*Trapelia*
18. Apothecia with a thalloid exciple	19
18. Apothecia with a proper exciple	21
19. Phycobiont *Nostoc*	*Pannaria*
19. Phycobiont green	20
20. Lower surface sparingly rhizinate	*Psoroma*
20. Lower surface without rhizines	*Squamarina*
21. Phycobiont blue-green	22
21. Phycobiont green	23
22. Upper surface with fine longitudinal striae, phycobiont *Scytonema*	*Coccocarpia*
22. Upper surface not striate, phycobiont *Nostoc*	*Parmeliella*
23. Squamules finely lacerate	*Phyllopsora*
23. Squamules not lacerate	29
24. Spores brown	25
24. Spores hyaline	26
25. Apothecia with thalloid exciple at maturity	*Rinodina*
25. Apothecia with proper exciple at maturity	*Diploicia*
26. Spores polarilocular	*Caloplaca*
26. Spores septate	28
27. Disk pink	*Baeomyces*
27. Disk black	*Toninia*
28. Phycobiont green *(Trebouxia)*	*Toninia*
28. Phycobiont blue-green *(Dichotrix)*	*Placynthium*
29. Lower surface sparingly rhizinate	*Psoromaria*
29. Lower surface without rhizines	*Psora*

The genus *Normandina* is never fertile.
Cladonia squamules are commonly found sterile.

Crustose thalli

Key to the artificial groups

1. Asci in fruiting structures which open by a pore (for example, perithecia)	Group A (page 23)
1. Asci in open fruiting structures, not opening by a pore (for example, apothecia)	2
2. Spores free in a mazaedium (not contained in asci) at maturity	Group B (page 25)
2. Spores retained in the ascus at maturity	3
3. Spores simple	Group C (page 26)
3. Spores septate or polarilocular	4
4. Spores once septate or polarilocular	Group D (page 27)
4. Spores more than once septate	5
5. Spores transversely septate only	Group E (page 28)
5. Spores muriform	Group F (page 31)

Key to genera of Group A: Fruiting structures opening by a pore

1. Spores simple	2
1. Spores septate	4
2. Fruiting structure an apothecium	*Pertusaria*
2. Fruiting structure a perithecium or perithecium-like structure	3
3. Asci with 1-8 spores	*Verrucaria*
3. Asci with many spores	*Trimmatothele*
4. Spores once septate	5
4. Spores more than once septate	14
5. Spores hyaline	6
5. Spores brown	11
6. Paraphyses absent or reticulately branched	7
6. Paraphyses simple or forked, but not reticulately branched	9
7. Fruiting structure usually compounded from several hymenia, opening by an irregular pore or slit	*Mycoporellum*
7. Fruiting structure of discrete perithecia, sometimes several immersed in thalline warts	8

8. Paraphyses soon gelatinizing *Thelidium*
8. Paraphyses permanent *Pleurotrema*

9. Fruiting body an apothecium immersed in warts *Ocellularia*
9. Fruiting body a perithecium 10

10. Thallus obligately foliicolous, always subcuticular (alga *Cephaleuros*) *Strigula*
10. Thallus not obligately foliicolous, but if on leaves supracuticular (alga *Trentepohlia* or *Phycopeltis*) *Porina*

11. The opening of the perithecium ornamented with a star-shaped disk *Asteroporum*
11. The opening of the perithecium simple or slit-like 12

12. Fruiting structure compounded from several hymenia opening by an irregular pore or slit *Mycoporellum*
12. Fruiting body with a single hymenium, opening by a pore 13

13. Spore walls unequally thickened, resulting in lenticular locules *Pyrenula*
13. Spore walls evenly thickened, locules cylindrical *Microthelia*

14. Spores transversely septate only 15
14. Spores muriform 27

15. Spores hyaline 16
15. Spores brown 25

16. Fruiting body an apothecium immersed in thalline warts *Ocellularia*
16. Fruiting body a perithecium, or group of perithecia, sometimes immersed in thalline warts 17

17. Fruiting body with several hymenia, opening by an irregular pore or by a slit *Mycoporellum*
17. Perithecium with a single hymenium opening by a pore 18

18. Perithecia immersed singly or in groups in stromatic tissue 19
18. Perithecia not aggregated in stromatic tissue 20

19. Spore walls unequally thickened, resulting in lenticular loculi *Trypethelium*
19. Spore walls of even thickness, locules cylindrical *Tomasellia*

20. Paraphyses branched and anastomosing 21
20. Paraphyses simple or forked towards the tips, but not anastomosing 23

21. Spore walls unequally thickened, resulting in lenticular locules *Pseudopyrenula*
21. Spore walls evenly thickened, locules cylindrical 22

22. Ostiolar canal markedly oblique or prostrate *Pleurotrema*
22. Ostiolar canal erect *Arthopyrenia*

23. Spore walls unequally thickened to produce lenticular locules, thallus not normally foliicolous *Pyrenula*
23. Spore walls thin, locules cylindrical 24

24. Perithecia with a corona of spreading bristles *Trichothelium*
24. Perithecia without a corona of spreading bristles 38

25. Fruiting structure usually compounded from several hymenia opening by an irregular pore or slit	*Mycoporellum*
25. Fruiting structure with a single hymenium, opening by an ostiole	26
26. Perithecia usually aggregated, immersed in stromatic tissue	*Melanotheca*
26. Perithecia single, not immersed in stromatic tissue	*Parathelium & Pyrenula*
27. Spores hyaline	28
27. Spores brown	33
28. Fruiting structure usually compounded from several hymenia, opening by an irregular pore or slit	*Dermatina*
28. Fruiting structure with a single hymenium opening by an ostiole	29
29. Paraphyses neither gelatinizing nor reticulately branched and anastomosing	30
29. Paraphyses gelatinizing or reticulately branched	31
30. Perithecium bearing a broad convex disk about the ostiole	*Aspidothelium*
30. Perithecium with an unornamented ostiole	*Clathroporina*
31. Perithecia immersed in stromatic tissue	*Laurera*
31. Perithecia not immersed in stromatic tissue	32
32. Ostiole extended into an oblique canal	*Campylothecium*
32. Ostiole straight, punctiform	*Polyblastiopsis*
33. Fruiting body usually compounded from several hymenia, opening by an irregular pore or slit	*Dermatina*
33. Fruiting body with a single hymenium, opening by a pore	34
34. Paraphyses simple, unbranched	35
34. Paraphyses reticulately branched and anastomosing	36
35. Perithecia grouped and immersed in stromatic tissue	*Bottaria*
35. Perithecia single, not immersed in stromatic tissue	*Anthracothecium*
36. Ostiolar canals erect	*Bottaria*
36. Ostiolar canals horizontal or oblique	37
37. Spores with lenticular or spherical locules	*Parmentaria*
37. Spores with cubical locules	*Pleurotheliopsis*
38. Thallus subcuticular	*Strigula*
38. Thallus supracuticular	*Porina*

Key to genera of Group B: spores free in a mazaedium at maturity

1. Spores simple	2
1. Spores septate	3
2. Spores pale (hyaline or yellow)	*Coniocybe*
2. Spores dark (brown)	*Chaenotheca*
3. Fruiting bodies sessile	4
3. Fruiting bodies at least shortly stipitate	5
4. Apothecium with a thalline margin	*Tylophoron*
4. Apothecium with a proper margin only	*Pyrgillus*

5. Spores 3-7 septate — *Stenocybe*
5. Spores 1 septate — *Calicium*

Key to genera of Group C: fruiting structures open, spores simple

1. Fruiting bodies stipitate on podetia or pseudopodetia — 2
1. Fruiting bodies sessile to immersed, but not stipitate — 6

    2. Podetia flattened laterally towards the top, apothecia facial on the flattened area — *Thysanothecium*
    2. Podetia not laterally flattened toward the top — 3

3. Apothecia terminal, conical or globose, hypothecium dark, podetia imperforate, rarely branched — 4
3. Apothecia terminal or lateral, never globose, hypothecium pale or if dark on much branched perforate pseudopodetia — 5

    4. Primary thallus granular — *Pilophorus*
    4. Primary thallus isidiate — *Gymnoderma*

5. Podetia richly di- or tri-chotomously branched — *Cladina*
5. Podetia little branched — *Cladonia*

    6. Phycobiont blue-green — 7
    6. Phycobiont green — 10

7. Thallus of discrete, short, thick, erect cylindrical lobes — *Peccania & Synalissa*
7. Thallus granular — 8

    8. Phycobiont filamentous *(Scytonema)* — *Porocyphus*
    8. Phycobiont coccoid — 9

9. Phycobiont *Gloeocapsa* (several cells clumped together) — *Pyrenopsis*
9. Phycobiont *Chroococcus* (cells single or in pairs) — *Pyrenopsidium*

    10. Spores 32 or more per ascus — 11
    10. Spores usually eight, but sometimes 1-16 per ascus — 16

11. Thallus yellow or orange-yellow — 12
11. Thallus other colours — 13

    12. Apothecia sessile or adnate — *Candelariella*
    12. Apothecia immersed — *Acarospora*

13. Apothecia with a thalloid exciple — 14
13. Apothecia with a proper exciple — 15

    14. Apothecia immersed — *Acarospora*
    14. Apothecia sessile to adnate — *Maronea*

15. Exciple pale and soft — *Biatorella*
15. Exciple black and hard — *Sarcogyne*

    16. Apothecia pierced by a well defined central pore — *Coccotrema*
    16. Apothecia not pierced by a central pore — 17

17. Spores brown — *Cyphelium*
17. Spores hyaline — 18

18. Apothecia immersed in warts — *Pertusaria*
18. Apothecia immersed or sessile on the thallus — 19

19. Thallus and/or apothecia yellow or orange-yellow — 20
19. Thallus and apothecia not yellow or orange-yellow — 22

20. Thallus or apothecia K + purple (parietin) — 21
20. Thallus and apothecia not K + purple (without parietin) — *Candelariella*

21. Thallus with distinct marginal lobes — *Fulgensia*
21. Thallus uniform — *Protoblastenia*

22. Paraphyses reticulate — 23
22. Paraphyses simple, branched or dissolving — 24

23. Apothecia immarginate — *Micarea*
23. Apothecia with a developed thalline exciple — *Ochrolechia*

24. Apothecia with a proper exciple — 25
24. Apothecia with a thalline exciple — 27

25. Thallus corticate or absent, not a loose hyphal weft — 29
25. Thallus a loose weft of hyphae — 26

26. Thallus with a thick black hypothallus — *Crocynia*
26. Thallus without a distinctive hypothallus — *Coenogonium*

27. Thallus with external cephalodia — *Placopsis*
27. Thallus without cephalodia — 28

28. Apothecia immersed — *Aspicilia*
28. Apothecia sessile — *Lecanora*

29. Apothecia well developed, round — *Lecidea*
29. Apothecia without an exciple or exciple poorly developed, irregular, minute. — *Xylographa*

Key to genera of Group D: fruiting structures open, spores once septate or polarilocular

1. Spores hyaline — 2
1. Spores brown — 21

2. Fruiting bodies raised on podetia — 3
2. Fruiting bodies sessile or immersed — 4

3. Apothecia terminal at maturity — *Baeomyces*
3. Apothecia lateral on flattened podetia at maturity — *Thysanothecium*

4. Spores polarilocular — 5
4. Spores truly septate — 6

5. Thallus sometimes areolate, apothecia laminal — *Caloplaca*
5. Thallus of areoles produced vertically to become minutely fruticose, apothecia terminal — *Polycaulonia*

6. Paraphyses reticulate — 7
6. Paraphyses unbranched or branched but not reticulate — 11

7. Hymenium elongate, between developed labia — *Opegrapha*
7. Hymenium irregular or ovoid but not with labia — 8

8. Asci globose	*Arthonia*
8. Asci clavate-cylindrical	9
9. Thallus with algae restricted to discrete minute warts or goniocysts	*Micarea*
9. Thallus with a more or less continuous algal layer	10
10. Hypothecium pale	*Megalospora*
10. Hypothecium dark	*Catinaria*
11. Spore walls unequally thickened, locules lenticular	12
11. Spore walls evenly thickened, locules cylindrical	13
12. Apothecia with well developed proper and thalloid margins	*Ocellularia*
12. Apothecia with poorly developed margins	*Phlyctella*
13. Apothecia with a thalloid margin at maturity	13
13. Apothecia with a proper margin at maturity	16
14. Thallus bright yellow or orange-yellow	*Candelariella*
14. Thallus of other colours	14
15. Spores 32 or more per ascus	*Maronea*
15. Spores 1-16 per ascus	*Phylctella*
16. Disk surrounded by a cobweb-like hyphal veil	*Byssoloma*
16. Disk without a veil	17
17. Margin of disk orange	*Dimerella*
17. Margin of disk not orange	18
18. Disk and margin pink	*Baeomyces*
18. Disk and margin pale or dark but not pink	19
19. Thallus thick, corticate	*Toninia*
19. Thallus thin, ecorticate	20
20. Spores small (less than 30μm long) with a thin septum	*Catillaria*
20. Spores large (more than 30 μm long) with a thick septum	*Megalospora*
21. Apothecia elongate, or minute and irregular	22
21. Apothecia more or less round	23
22. Spore walls unequally thickened, locules lenticular	*Phaeotrema*
22. Spore walls evenly thickened, locules cylindrical	*Melaspilea*
23. Apothecia with a thalloid margin at maturity	*Rinodina*
23. Apothecia with a proper margin at maturity	*Buellia*

Key to genera of Group E: fruiting structure open, spores transversely two or more septate

1. Spores brown	2
1. Spores hyaline	9
2. Apothecia round, well developed and not immersed in stromatic tissue	3
2. Apothecia elongate, stellate, irregular and immersed in stromatic tissue or rudimentary	4
3. Apothecia with a thalloid margin at maturity	*Rinodina*
3. Apothecia without a thalloid margin at maturity	*Buellia*

4. Apothecia immersed in stromatic tissue	5
4. Apothecia not immersed in stromatic tissue	6
5. Apothecia one per stromatic wart	*Phaeotrema*
5. Apothecia commonly several per stromatic wart	*Sarcographa*
6. Apothecia more or less closed by a proper margin which disintegrates to leave a thalloid margin only	*Phaeotrema*
6. Apothecia rudimentary or with a persistent proper margin	7
7. Apothecia only rudimentary, spore walls evenly thickened, spore locules cylindrical	*Gymnographa*
7. Apothecia with a well developed proper margin, and sometimes a thalloid margin also, spore walls unequally thickened to produce lenticular locules, or wall uniformly thin	8
8. Exciple of distinct labia, not entire	*Phaeographis*
8. Exciple continuous, entire	*Dirinastrum*
9. Fruiting bodies stipitate	10
9. Fruiting bodies sessile to immersed	11
10. Spores 1-3 septate, elongate	*Baeomyces*
10. Spores up to 100 septate, filiform	*Gomphyllus*
11. Apothecia grouped in stromatic tissue	12
11. Apothecia not grouped in stromatic tissue	15
12. Paraphyses simple, spore walls even in thickness, spore locules cylindrical	*Glyphis*
12. Paraphyses branched and anastomosing, spore walls unevenly thickened to form lenticular loculi	13
13. Thallus foliicolous, alga *Phycopeltis*	*Mazosia*
13. Thallus not foliicolous, alga *Trentepohlia*	*14*
14. Hypothecium pale	*Enterographa*
14. Hypothecium dark	*Chiodecton*
15. Apothecia elongate or stellately irregular; if ovoid, exciple poorly developed or absent	16
15. Apothecia round, exciple usually well developed	18
16. Twin hymenia parallel to the axis of the apothecium	*Diplogramma*
16. Single hymenium only	17
17. Paraphyses reticulately branched and anastomosing	18
17. Paraphyses simple or forked, but not anastomosing	20
18. Apothecia with a well developed margin	19
18. Apothecia without a margin	21
19. Spore walls unequally thickened to produce lenticular locules	*Helminthocarpon*
19. Spore walls of even thickness, locules cylindrical	25
20. Spore walls of even thickness, locules cylindrical	23
20. Spore walls unequally thickened, locules lenticular	*Graphis*
21. Thallus foliicolous	47
21. Thallus not foliicolous	22

22. Spores with asymmetric locules	*Arthonia*
22. Spores with symmetrical locules	*Melaspilea*
23. Exciple lateral along the margin of the hymenium only	*Aulaxina*
23. Margin either surrounding the apothecium or absent	24
24. Hypothecium black, carbonaceous	*Lecanactis*
24. Hypothecium pale	*Melampydium*
25. Exciple lateral along the margin of the hymenium only	*Opegrapha*
25. Exciple surrounding the roundish hymenium, or absent	26
26. Exciple carbonaceous, without thalloid covering	*Lecanactis*
26. Exciple usually absent, but if present with a thalloid covering	27
27. Thallus with a uniform, continuous algal layer	*Schismatomma*
27. Thallus with algae restricted to small tubercles (goniocysts)	*Micarea*
28. Paraphyses reticulately branched and anastomosing	29
28. Paraphyses simple or forked, but not anastomosing	32
29. Upper surface of thallus covered by black hairs	*Tricharia*
29. Upper surface of thallus naked	30
30. Exciple absent, or if present poorly developed with a thalloid cover	48
30. Exciple well developed, without thalloid cover	31
31. Spores acicular, transversely septate only, Phycobiont *Trentepohlia* on wood and rock	*Lecanactis*
31. Spores ellipsoid or oblong-ellipsoid, transversely septate or muriform, Phycobiont *Trebouxia*, foliicolous	*Tapellaria*
32. Apothecia with a thalloid margin at maturity	33
32. Apothecia without a thalloid margin at maturity	39
33. Spore walls unequally thickened, locules lenticular	34
33. Spore walls of even thickness, locules cylindrical	35
34. Disk initially enclosed by a proper exciple which later cracks away	*Ocellularia*
34. Disk not enclosed by a proper exciple, although a poorly developed proper exciple may occur	*Phlyctella*
35. Hypothecium carbonaceous	*Sagiolechia*
35. Hypothecium pale	36
36. Spores oblong-ellipsoid, 1-3 or rarely more septate	37
36. Spores acicular 5+ septate	*Haematomma*
37. Phycobiont *Trentepohlia*, forming chains in the medulla	*Gyalecta*
37. Phycobiont Trebouxoid, not forming chains	38
38. Apothecium pink	*Baeomyces*
38. Apothecium pale or dark but not pink	*Lecania*
39. Apothecial margin with an arachnoid veil	*Byssoloma*
39. Apothecial margin without a veil	41

40. Phycobiont green — 46
40. Phycobiont blue-green — 42

41. Phycobiont Trebouxoid, not in chains — 43
41. Phycobiont *Trentepohlia*, forming chains in the medulla — 44

42. Spore walls unequally thickened, locules lenticular — *Bombyliospora*
42. Spore walls evenly thickened, locules cylindrical — 43

43. Spores many (up to 100) septate — *Conotremopsis*
43. Spores few septate — *Bacidia*

44. Hypothecium pale — *Gyalecta*
44. Hypothecium carbonaceous — 45

45. Spore walls unequally thickened, locules lenticular — *Lecanactis*
45. Spore walls of even thickness, locules cylindrical — *Sagiolechia*

46. Phycobiont *Nostoc*, over mosses — *Arctomia*
46. Phycobiont *Dichothrix*, on soil — *Placynthium*

47. Phycobiont coccoid (*Trebouxia* like) — *Echinoplaca*
47. Phycobiont plate forming (*Phycopeltis*) — *Mazosia*

48. Exciple poorly developed, hyothecium dark — *Schismatomma*
48. Exciple well developed, hypothecium pale — *Haematomma*

Key to genera of Group F: fruiting structures open, spores muriform

1. Spores brown — 2
1. Spores hyaline — 12

2. Paraphyses reticulately branched and anastomosing — 3
2. Paraphyses simple or forked, but not anastomosing — 5

3. Apothecia immersed deep in stromatic tissue — *Tremotylium*
3. Apothecia sessile or slightly immersed, but not in stromatic tissue — 4

4. Thallus rudimentary, asci pyriform or globose — *Arthothelium*
4. Thallus well developed, areolate, asci clavate — *Rhizocarpon*

5. Fruiting bodies deeply immersed in the thallus or in stromatic tissue, often appearing punctiform — 6
5. Fruiting bodies sessile to adnate or slightly immersed — 9

6. Apothecia grouped in stromatic patches — *Sarcographina*
6. Apothecia not grouped in stromatic patches — 7

7. Hymenium markedly elongate to stellate — *Phaeographina*
7. Hymenium round to ovoid — 8

8. The proper exciple disintegrating to leave the hymenium free within the thalloid exciple — *Leptotrema*
8. The proper exciple retained, surrounded by a thalloid exciple also — *Diploschistes*

9. Apothecium markedly elongate to stellate — *Phaeographina*
9. Apothecia round to ovate — 10

10. Apothecia immersed or with a distinct thalloid exciple — *Diploschistes*
10. Apothecia with a proper exciple only — 11

11. Paraphyses unbranched, tightly packed, spores not mucilagenous — *Buellia*
11. Paraphyses reticulately branched and anastomosing, loose, spores mucilagenous — *Rhizocarpon*

12. Paraphyses branched and anastomosing — 13
12. Paraphyses simple or branched, but not anastomosing — 19

13. Thallus foliicolous — 14
13. Thallus not foliicolous — 17

14. Exciple well developed — *Tricharia*
14. Exciple poorly developed — 15

15. Epithecium without algae — *Gyalectidium*
15. Epithecium with algae — 16

16. Thallus corticate — *Echinoplaca*
16. Thallus ecorticate — *Sporopodium*

17. Thalloid exciple well developed — *Tremotylium*
17. Thalloid exciple absent — 18

18. With a well developed proper exciple, asci clavate — *Rhizocarpon*
18. Without an exciple, asci pyriform or globular — *Arthothelium*

19. Fruiting bodies markedly elongate, or grouped in raised stromatic patches — 20
19. Fruiting bodies more or less circular, not grouped in stromatic patches — 23

20. Fruiting bodies markedly elongate, with lateral exciple only — *Graphina*
20. Fruiting bodies irregularly ovoid to round, immersed in stromatic patches — 21

21. Spore walls of even thickness, locules cubical — *Enterodictyon*
21. Spore walls unequally thickened, locules spherical to lenticular — 22

22. Apothecia grouped in stromatic patches — *Medusulina*
22. Apothecia single in warts — *Thelotrema*

23. Spore walls of even thickness, locules cubical — 24
23. Spore walls unequally thickened, locules spherical to lenticular — *Thelotrema*

24. Spores with a distinct mucilagenous covering — *Rhizocarpon*
24. Spores without a mucilagenous covering — 25

25. Exciple never well developed, thallus obligately foliicolous — *Gyalectidium*
25. Exciple usually well developed, not obligately foliicolous — 25

26. Phycobiont *Trentepohlia*, algae beneath the hymenium — *Gyalecta*
26. Phycobiont coccoid, without algae beneath the hymenium — *Lopadium*

Filamentous thalli

1.	Thallus of algal filaments ensheathed in fungal hyphae	2
1.	Thallus a loose weft of hyphae loosely entangling algal cells or filaments	4
2.	Algal filaments blue-green *(Stigonema)*	3
2.	Algal filaments green *(Trentepohlia)*	*Cystocoleus*
3.	Apothecia immersed in swellings on filaments	*Ephebe*
3.	Apothecia terminal on filaments	*Spilonema*
4.	Spores septate (up to 100)	*Conotremopsis*
4.	Spores simple or 1 septate	5
5.	Thallus with thick black hypothallus	*Crocynia*
5.	Thallus without a distinctive hypothallus	6
6.	Thallus bright yellow	*Chrysothrix*
6.	Thallus white to grey	*Coenogonium*

Leprose thalli

1. Thallus crustose 2
1. Thallus fruticose *Leprocaulon*
    2. Thallus white or grey to green *Lepraria*
    2. Thallus bright yellow *Chrysothrix*

Basidiolichens

1. Thallus shelf-like, partly free from the substrate, blue-black to deep green *Dictyonema*
1. Thallus prostrate, adnate to substrate, blood red at least on the margins *Herpothallon*

Descriptions of Lichen Genera

Under the heading Morphology for each genus the description first deals with the form of the thallus, its anatomy and mode of attachment to the substrate. In the second sentence the reproductive structures are dealt with in order fruiting body form, ascus form and spore form. The third sentence deals with pycnidial structure.

The second heading, Chemistry, is followed by a figure in brackets, which indicates the number of species on which the follow data compilation is based. This is followed by a list of the secondary lichen chemical products reported for the genus in diminishing order of frequency, with the number of species having each constituent indicated in brackets.

The phycobiont is then indicated at least tentatively for all genera.

Notes on the distribution are necessarily sketchy. The number of species reported refers to all records for Australia (including Tasmania) and is followed by a number in brackets which is an estimate of the number of species in the genus for the world.

ACAROSPORA MASS.: *Ric. Lich. Crost.* 27 (1852)

MORPHOLOGY: *Thallus* crustose to squamulose, areoles or squamules more or less uniform or forming marginal lobes; attached to the substrate only by medullary hyphae; differentiated with a plectenchymatous upper and sometimes lower cortex; medulla compact. *Ascocarp* an apothecium, immersed or rarely sessile, single or many per areole, disk round or irregular, thalloid exciple; hypothecium pale, over an algal layer; paraphyses unbranched, septate; asci unitunicate, thick walled, inflated, I+ pale blue, tholus I+ pale blue, many spored; ascospores minute, simple, hyaline. *Pycnidia* immersed; fulcrum exobasidial; pycnospores oblong-ellipsoid to globose.
CHEMISTRY: (39 species) rhizocarpic acid (13), gyrophoric acid (8), acaronoic acid (4), lecanoric acid (4), norstictic acid (2), usnic acid (1).
PHYCOBIONT: *?Trebouxia.*
DISTRIBUTION: Widespread on rock and soil, especially on drier sites in all mainland states. 15 species reported (300).

ALECTORIA ACH.: *Lich. Univ.* 592 (1810)

MORPHOLOGY: *Thallus* fruticose, markedly elongate, prostrate or erect; attached by a basal holdfast; lobes terete or flattened, rarely angled; cortex horny, of agglutinated longitudinal hyphae; medulla of loose arachnoid longitudinal hyphae, distinctly segmented; pseudocyphellae and soredia not uncommon. *Ascocarp* an apothecium, lateral, stalked, thalloid margin with or without cilia; disk concave, brown to black; hypothecium pale, over an algal layer; paraphyses branched and anastomosing, asci unitunicate, thick walled, I+, tholus I+ blue, 4-8 spored; ascospores simple, ellipsoid, hyaline to brown, thin walled. *Pycnidia* immersed in thalloid warts; fulcrum endobasidial; pycnospores bacilliform.
CHEMISTRY: (8 species) usnic acid (8), alectoronic acid (6), thamnolic acid (5), barbatic acid (4), olivetoric acid (3), squamatic acid (3), α-collatolic acid (2), alectorialic acid (1), barbatolic acid (1), diffractaic acid (1).
PHYCOBIONT: *Trebouxia.*
DISTRIBUTION: On bark and soil, restricted to sub-alpine and alpine areas in Tasmania and New South Wales. 3 species reported (40).

ANAPTYCHIA KORB.: *Grund. Krypt.* 197 (1848)

MORPHOLOGY: *Thallus* foliose to more or less fruticose, usually somewhat ascending, attached to the substrate by rhizines; the lobes usually only a few millimetres broad, often

ciliate; differentiated into upper cortex, medulla and lower cortex; upper cortex of agglutinated longitudinal hyphae, not pseudoparenchymatous; medulla woolly of thin walled hyphae, lower cortex may be almost identical to upper, or less developed. *Ascocarp* an apothecium, laminal or terminal; exciple thalloid; disk dark, often pruinose; paraphyses unbranched; asci clavate unitunicate, thick walled, I+ blue, tholus I+ blue, 8 spored; ascospores 1 septate, without blastidia, walls thickened, dark at maturity. *Pycnidia* immersed or protruding; fulcrum endobasidial; pycnospores bacilliform.
CHEMISTRY: (6 species) atranorin (6), zeorin (6).
PHYCOBIONT: Trebouxia.
DISTRIBUTION: On trees and rocks, in wetter parts of Queensland, New South Wales, Victoria and Tasmania. 2 species reported (9).

ANTHRACOTHECIUM HAMPE ex MASS.: *Atti Ist. veneto Sci. ser 3,* 5: 330 (1860)

MORPHOLOGY: Thallus crustose, epi- or endophloic, not differentiated into layers; uniform. *Ascocarp* a perithecium, scattered, or grouped but not aggregate in a stroma; ostiole erect; paraphyses unbranched, free; asci clavate, bitunicate, endoascus well developed, I−, 8 spored; ascospores brown, ellipsoid to oblong, muriform, the walls unequally thickened to produce lenticular or rhomboid lumina. *Pycnidia* immersed; fulcrum exobasidial; pycnospores elongate-cylindrical to filiform.
CHEMISTRY: (1 species) fallacinal (1), parietin (1).
PHYCOBIONT: ?Trentepohlia.
DISTRIBUTION: On bark, in wetter areas of Queensland and Victoria. 17 species reported (93).

ANZIA STIZ.: *Flora* 44, 390 (1861)

MORPHOLOGY: Thallus foliose, finely divided with more or less articulated lobes; upper surface grey, corticate, of anticlinal hyphae; lower surface a spongy tomentum of usually dark anastomosing medullary hyphae; some rhizines coming direct from the medulla. *Ascocarp* an apothecium, laminal, circular, stipitate; margin thalline; hypothecium pale; paraphyses unbranched or little branched; asci cylindrical, unitunicate, I+, tholus large and I+, many spored; ascospores small, bicornute, simple, hyaline, thin walled. *Pycnidia* immersed; fulcrum exobasidial; pycnospores short-cylindrical or curved.
CHEMISTRY: (8 species) atranorin (7), divaricatic acid (7), lobaric acid (2), sekikaic acid (2), anziaic acid (1), caperatic acid (1), chloroatranorin (1), fumarprotocetraric acid (1).
PHYCOBIONT: Trebouxia.
DISTRIBUTION: New South Wales and Victoria, on bark. 1 species reported (28).

ARCTOMIA TH. FR.: *Nova Acta Reg. Soc. Sci. Upsal.* 3, 3, 387 (1861)

MORPHOLOGY: Thallus crustose, granular to somewhat lobed; with rhizines; upper surface with one or two cortical cell layers. *Ascocarp* an apothecium, sessile, with a proper exciple only; paraphyses unbranched or reticulate with thickened ends; hypothecium pale; asci cylindrical-clavate, unitunicate, I+, thick walled, usually 8 spored; ascospores transversely 5-7 septate, hyaline. *Pycnidia* immersed; fulcrum endobasidial, pycnospores bacilliform.
CHEMISTRY: No data.
PHYCOBIONT: Nostoc.
DISTRIBUTION: On mosses, in Tasmania. 1 species reported (2).

ARTHONIA ACH.: *Neues J. Bot.* 1(3): 3 (1806)

MORPHOLOGY: Thallus crustose ecorticate uniform, partly or wholly immersed in the substrate. *Ascocarp* a round to lirreliform pseudothecium, sessile to immersed in the thallus, without an exciple, disk flat to convex, usually dark; hypothecium usually dark; paraphyses branched and anastomosing, indistinct; asci globose to pyriform or obovoid, bitunicate, I—, 8 spored; ascospores oblong, ovoid or asymmetric, transversely septate, often with very unequal locules, usually hyaline but rarely brown. *Pycnidia* laminal; fulcrum exobasidial; pycnospores elongate cylindrical; stylospores also formed.
CHEMISTRY: (4 species) arthoniaic acid (1), emodin (1), fallacinal (1), lepranthin (1), rocellaric acid (1), teloschistin (1).
PHYCOBIONT: Phycopeltis, Trentepohlia, ?Pleurococcus.
DISTRIBUTION: On trees, rocks and soil, probably throughout Australia and Tasmania but especially in warmer and wetter areas. 28 species reported (500).

ARTHOPYRENIA MASS: *Ric. Lich. Crost.* 165 (1852)

MORPHOLOGY: Thallus crustose, ecorticate, uniform, undifferentiated, usually immersed in the substrate. *Ascocarp* a perithecium-like pseudothecium, opening by a punctiform erect ostiole; paraphyses branched and anastomosing often gelatinizing; asci bitunicate, I—, 8 spored; ascospores asymmetric or oval to oblong with or without terminal protruberances, transversely 1-5 septate, the cells cylindrical, walls hyaline. *Pycnidia* fulcrum endobasidial; pycnospores bacilliform; 1 septate stylospores also formed.
CHEMISTRY: No data.
PHYCOBIONT: Hyella.
DISTRIBUTION: Queensland, New South Wales and Victoria. 17 species reported (200).

ARTHOTHELIUM MASS.: *Ric. Lich. Crost.* 54 (1852)

MORPHOLOGY: Thallus crustose, ecorticate, uniform, partly or wholly immersed in the substrate. *Ascocarp* a round to lirreliform pseudothecium, sessile to immersed in the thallus, without an exciple, disk flat to convex, usually dark; hypothecium usually dark, paraphyses branched and anastomosing, indistinct; asci globose to pyriform or obovoid, bitunicate, I—, 8 spored; ascospores oblong to ovoid, muriform with angular locules, usually hyaline, sometimes brown. *Pycnidia* laminal; fulcrum exobasidial; pycnospores elongate cylindrical; stylospores also formed.
CHEMISTRY: No data.
PHYCOBIONT: Trentepohlia.
DISTRIBUTION: Queensland, New South Wales, Victoria and Tasmania. 11 species reported (80).

ASPICILIA MASS.: *Ric. Lich. Crost.* 169 (1852)

MORPHOLOGY: Thallus crustose to areolate, foliose or fruticose, attached by hyphae, upper cortex often distinctly paraplectenchymatous. *Ascocarp* an apothecium, immersed in the thallus, with a thalloid exciple; paraphyses unbranched, often moniliform; hypothecium hyaline; asci unitunicate, thin walled I+ blue, tholus I+ blue, flaccid, up to 8 (but often fewer) spores; ascospores rather large, thin walled, simple, hyaline. *Pycnidia* immersed; fulcrum exobasidial; pycnospores cylindrical or filiform, straight or curved.

CHEMISTRY: (35 species) norstictic acid (11), aspicilin (5), unknown phenolics (5), atranorin (2), physodalic acid (2), gyrophoric acid (1), placodin (1), taurine (1).
PHYCOBIONT: *Trebouxia.*
DISTRIBUTION: Common on rocks and soil, in alpine and dry areas of Australia. 5 species reported (100).

ASPIDOTHELIUM VAIN. em R. SANT.: *Symb. bot. Upsal.* 12: 279 (1952)

MORPHOLOGY: *Thallus* crustose, uniform, ecorticate. *Ascocarp* a perithecium, solitary, opening by an erect ostiole which extends into a canal with a broad light coloured flange around its opening; hymenium without algae; paraphyses thin, unbranched; asci oblong to oblong-cylindrical, bitunicate, thick walled, I−, 2-8 spored; ascospores oblong to spindle-shaped with terminal protruberances, muriform, hyaline. *Pycnidia* sessile or immersed; fulcrum endobasidial; pycnospores bacilliform.
CHEMISTRY: No data.
PHYCOBIONT: *Chlorococcus*-like.
DISTRIBUTION: Queensland, on leaves. 1 species reported (4).

ASTEROPORUM MULL. ARG.: *Rev. Mycol.* 6: 14 (1884)

MORPHOLOGY: *Thallus* crustose, epi- or endophloic, uniform, undifferentiated. *Ascocarp* a perithecium-like pseudothecium, single, not aggregated in stroma, sessile or partly immersed, the ostiole erect with a star-shaped appendage; paraphyses branched and reticulate; asci bitunicate, I−, oblong to cylindrical, 8 spored; ascospores 1 septate with unequal locules, brown. *Pycnidia* not known.
CHEMISTRY: No data.
PHYCOBIONT: *Trentepohlia*-like.
DISTRIBUTION: On bark, in Queensland and New South Wales. 2 species reported (3).

AULAXINA FEE.: *Essai Crypt.* 60 (1824)

MORPHOLOGY: *Thallus* crustose, attached by rhizoidal hyphae, undifferentiated, uniform, ecorticate. *Ascocarp* a lirelliform apothecium; with a carbonaceaous, proper exciple lateral to the hymenium; hymenium linear, unbranched; hypothecium light brown; paraphyses thin, reticulate; asci oblong to spherical, unitunicate, I−, thin walled, 2-8 spored; ascospores transversely 3-8 septate, elongate, hyaline, I+ violet, walls thin. *Pycnidia* adnate; fulcrum endobasidial, pycnospores bacilliform.
CHEMISTRY: No data.
PHYCOBIONT: *Palmella*-like.
DISTRIBUTION: On leaves, in Queensland. 1 species reported (8).

BACIDIA DE NOT.: *Giorn. Bot. Ital.* 2: 189 (1846)

MORPHOLOGY: *Thallus* crustose, attached by rhizoidal hyphae, uniform, ecorticate. *Ascocarp* an apothecium, sessile to somewhat immersed in the thallus; disk round, flat to convex; hypothecium pale to dark; paraphyses unbranched, often capitate; asci long-clavate to cylindrical, unitunicate, I + blue, 8-16 spored; ascospores with 3-many transverse septa, fusiform to acicular, both ends alike or one end pointed one convex, curved, straight or

helically interwoven, walls thin, transverse septa often difficult to see. *Pycnidia* immersed; fulcrum exobasidial; pycnospores bacilliform to acicular.
CHEMISTRY: (14 species) atranorin (2), rhizocarpic acid (1).
PHYCOBIONT: *Trebouxia.*
DISTRIBUTION: Queensland, New South Wales, Victoria and Tasmania. 38 species reported (400).

BAEOMYCES PERS.: *Ann. Bot. (Usteri)* 7, 19 (1794)

MORPHOLOGY: *Thallus* crustose, granular or sub-squamulose; attached by medullary hyphae, ecorticate. *Ascocarp* an apothecium terminal on an erect unbranched hollow stipe; the stipe naked or with a thalloid covering; disk flat to almost spherical, pale or often pink; hypothecium usually pale; paraphyses unbranched; asci narrow, unitunicate, tholus I+ blue, 8 spored; ascospores ellipsoid to fusiform, simple to 3 septate, walls thin, hyaline. *Pycnidia* immersed in thalloid warts; fulcrum exobasidial; pycnospores bacilliform.
CHEMISTRY: (9 species) norstictic acid (5), baeomycesic acid (3), squamatic acid (2), stictic acid (2), atranorin (1).
PHYCOBIONT: *Coccomyxa.*
DISTRIBUTION: On acid soils or rocks, in Queensland, Victoria and Tasmania. 7 species reported (44).

BIATORELLA TH.FR.: *Nova Acta R. Soc. Scient. Upsal.* (Ser 3), 3: 299 (1861)

MORPHOLOGY: *Thallus* crustose, uniform or with marginal lobes, ecorticate or with a poorly developed upper cortex; attached by medullary or prothallial hyphae. *Ascocarp* an apothecium, sessile; disk circular, pale to flesh coloured, margin proper; hypothecium pale; paraphyses rarely branched, free; asci thick walled, unitunicate, I+ pale blue, tholus I+ pale blue, many spored; ascospores ellipsoid to globose, simple, hyaline, minute, thin walled. *Pycnidia* immersed in the thallus or in small warts; fulcrum exobasidial; pycnospores short-cylindrical.
CHEMISTRY: (4 species) parietin (2).
PHYCOBIONT: *?Myrmecia, ?Trebouxia.*
DISTRIBUTION: Queensland and Victoria. 4 species reported (50).

BOMBYLIOSPORA MASS.: *Ric. Lich. Crost.* 114 (1852)

MORPHOLOGY: *Thallus* crustose, undifferentiated, uniform, attached by medullary or prothallial hyphae, ecorticate. *Ascocarp* an apothecium, sessile to adnate; disk round, flat to slightly convex; hypothecium pale or dark, devoid of algae; margin proper; paraphyses unbranched, thread-like; asci broadly clavate unitunicate I+ blue, 1-8 spored; ascospores transversely 3-11 septate, hyaline (or pale brown), the septa unequally thickened to give lenticular loculi. *Pycnidia* immersed; fulcrum exobasidial; pycnospores bacilliform.
CHEMISTRY: no data.
PHYCOBIONT: *Trebouxia*-like.
DISTRIBUTION: Queensland, New South Wales, Victoria, Tasmania and South Australia. 10 species reported (30).

BOTTARIA MASS.: *Misc. Lich.* 42 (1856b)

MORPHOLOGY: Thallus crustose, epi- or endophloic, uniform, ecorticate, or with an amorphous upper cortex. *Ascocarp* a perithecium; perithecia solitary or grouped in a stroma, globose with vertical ostioles; paraphyses unbranched or reticulately branched; asci cylindrical-clavate, bitunicate, I−, 1-8 spored; ascospores muriform, brown, with thin walled septa. *Pycnidia* small; fulcrum exobasidial; pycnospores filiform.
CHEMISTRY: No data.
PHYCOBIONT: Trentepohlia.
DISTRIBUTION: Queensland. 2 species reported (6).

BUELLIA DE NOT.: *Giorn. Bot. Ital.* 2(1): 195 (1846)

MORPHOLOGY: Thallus crustose, sometimes distinctly lobed at the margins; or squamulose, sometimes endolithic; usually ecorticate but sometimes with a well developed cortex; attached by medullary or prothallial hyphae. *Ascocarp* an apothecium; immersed, adnate or sessile; disk round, margin proper; hypothecium dark, without algae; paraphyses unbranched, thickened at the tips; asci clavate or cylindrical, unitunicate, I+ blue, tholus I+ blue, usually 8 (rarely 32) spored; ascospores usually 1 septate, sometimes falsely 4 celled by ingrowth of outer wall, wall and septa sometimes very unequally thickened, wall brown. *Pycnidia* immersed or in warts; fulcrum endobasidial; pycnospores bacilliform.
CHEMISTRY: (59 species) atranorin (19), norstictic acid (17), unknown phenolics (8), arthothelin (6), 2-7 dichloronorlichexanthone (6), 2-chloronorlichexanthone (5), thiophanic acid (4), granulosin (2), unidentified xanthones (2), chloroatranorin (1), constictic acid (1), fumarprotocetraric acid (1), physodalic acid (1), stictic acid (1).
PHYCOBIONT: Trebouxia.
DISTRIBUTION: On rock, bark and wood, throughout the region. 55 species reported (600).

BULBOTHRIX HALE: *Phytologia* 28: 479 (1974b)

MORPHOLOGY: Thallus foliose; attached to the substrate by rhizines; with well developed upper and lower cortices, upper cortex of pallisade plectenchyma; lower surface pale to black, rhizines simple to richly dichotomously branched; margins of lobes with bulbate cilia. *Ascocarp* an apothecium; sessile; margin thalloid, with cilia or corona of sessile pycnidia; disk round, flat or concave; hypothecium pale, with algae; paraphyses unbranched, capitate; asci cylindrical-clavate, unitunicate, I+ blue, 8 spored; ascospores simple, ovoid or bicornute, thick walled. *Pycnidia* immersed; fulcrum endobasidial, bayonet-like; pycnospores bacilliform.
CHEMISTRY: (31 species) atranorin (31), salazinic acid (9), gyrophoric acid (7), lecanoric acid (3), norstictic acid (2), protocetraric acid (2), lobaric acid (1), norlobaridone (1), skyrin (1).
PHYCOBIONT: Trebouxia.
DISTRIBUTION: On trees, in coastal Queensland and New South Wales. 5 species reported (31).

BYSSOLOMA TREVIS.: *Spighe e Paglie,* 6 (1853)

MORPHOLOGY: Thallus crustose, ecorticate, foliicolous. *Ascocarp* an apothecium, sessile or adnate; disk round, flat or slightly convex, margin proper; hypothecium dark; paraphyses unbranched or little branched; asci cylindrical-clavate, unitunicate, tholus I+ blue, 8 spored;

ascospores transversely 1-3 septate, hyaline. *Pycnidia* sessile, globose; fulcrum endobasidial; pycnospores spherical to bacilliform.
CHEMISTRY: (1 species) 2-chloroemodin, emodin, fragilin.
PHYCOBIONT: *Trebouxia*-like.
DISTRIBUTION: Queensland and New South Wales. 1 species reported (6).

CALICIUM PERS.: *Ann. Bot. (Usteri)* 7: 20 (1794)

MORPHOLOGY: *Thallus* crustose, epi- or endophloic, uniform, ecorticate. *Ascocarp* a minutely (less than 2 mm tall) stipitate apothecium; usually cupular; disk flat to strongly convex, circular; margin and stipe without algae; hypothecium brown; paraphyses thread-like; asci thin walled, unitunicate (disintegrating into a mazaedium of ascal and hyphal fragments with spores) 8 spored; ascospores 1 septate, brown to black. *Pycnidia* immersed; fulcrum exobasidial; pycnospores bacilliform.
CHEMISTRY: (3 species) vulpinic acid (3), atranorin (1), norstictic acid (1).
PHYCOBIONT: *Chlorella, Stichoccoccus.*
DISTRIBUTION: Queensland, Victoria, Tasmania and South Australia. 36 species reported (100).

CALOPLACA TH. FR.: *Lich. Scand.* 1: 167 (1871)

MORPHOLOGY: *Thallus* crustose; uniform to lobed at the margins; attached by medullary or prothallial hyphae; ecorticate or with a well developed upper cortex; cortex often paraplectenchymatous, of vertical agglutinate hyphae. *Ascocarp* an apothecium, immersed to sessile; disk round, with parietin; margin proper or thalloid; hypothecium pale, with or without algae; paraphyses unbranched, thickened at the tips; asci clavate, unitunicate, I+ blue, tholus I+ blue, 8 spored; ascospores polarilocular, with unequally thickened walls but the two loculi linked by a canal. *Pycnidia* immersed; fulcrum endobasidial; pycnospores bacilliform.
CHEMISTRY: (of many known 110 considered) emodin (87), parietin (87), fallacinal (71), erythroglaucin (66), teloschistin (66), xanthorin (65), 2-chloroemodin (17), fragilin (15), parietinic acid (15), 1-0-methyl-fragilin (2), usnic acid (2), atranorin (1), citeorosein (1), emodic acid (1), emodinal (1), gyrophoric acid (1), lichexanthone (1).
PHYCOBIONT: *Trebouxia.*
DISTRIBUTION: On rock, bark and soil, in all states. 26 species reported (450).

CAMPYLOTHECIUM MULL. ARG.: *Flora, Jena* 66, 245 (1883)

MORPHOLOGY: *Thallus* crustose, without lobes, epi-or endophloic, uniform, ecorticate or with a cortex of horizontal hyphae. *Ascocarp* a perithecium opening by more or less oblique ostiolar canal, usually clad in thalline material; paraphyses reticulate; asci cylindrical-clavate, bitunicate, usually thick walled, I−, 1-8 spored; ascospores muriform, hyaline, thin walled, locules angular. *Pycnidia* not known.
CHEMISTRY: No data.
PHYCOBIONT: *Trentepohlia*-like.
DISTRIBUTION: Queensland. 2 species reported (9).

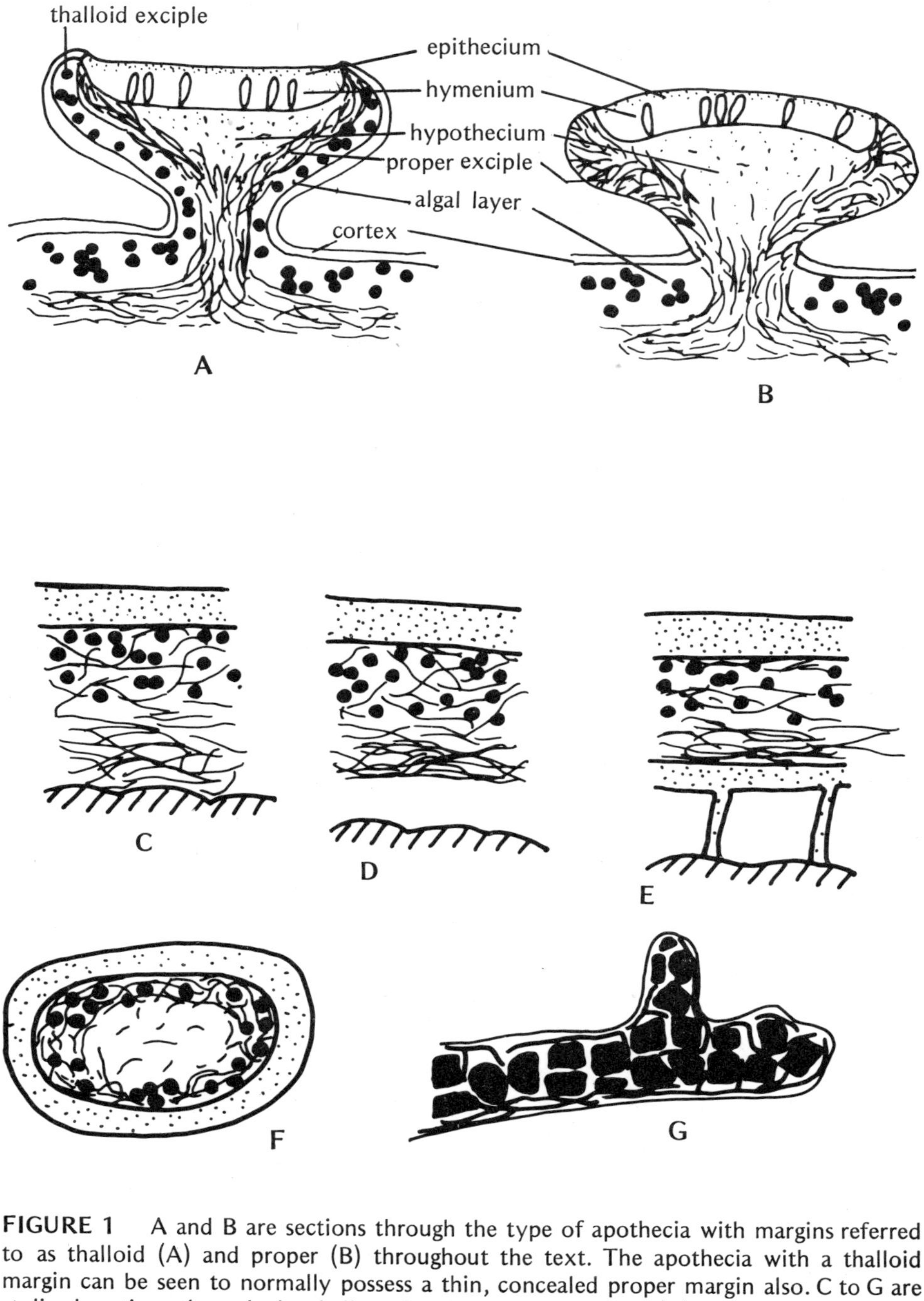

FIGURE 1 A and B are sections through the type of apothecia with margins referred to as thalloid (A) and proper (B) throughout the text. The apothecia with a thalloid margin can be seen to normally possess a thin, concealed proper margin also. C to G are stylized sections through the thallus of various growth forms: (C) crustose, (D) squamulose, (E) foliose, (F) fruticose, and (G) filamentous. Stippled areas represent cortex, hatched substrate; algae are represented by black dots or patches, and loosely woven hyphae by lines. In the filamentous form the hyphae are within a layer of algal gel.

CANDELARIA MASS.: *Ric. Lich. Crost.* 567 (1852)

MORPHOLOGY: Thallus foliose, much lacerate, coloured yellow with pulvinic acid derivatives, attached by rhizoids, upper and lower cortices well developed, paraplectenchymatous, lower cortex rhizinate. *Ascocarp* an apothecium, sessile, round, margin thalline; disk depressed, yellow; asci clavate, unitunicate, thick walled I+ blue, tholus I+ blue; 16-32 spored; paraphyses little branched; ascospores simple, often with two large oil drops, or rarely 1 septate, hyaline. *Pycnidia* immersed in small warts; fulcrum exobasidial; pycnospores ellipsoid.
CHEMISTRY: (2 species) calycin (2), pulvinic dilactone (2).
PHYCOBIONT: ?Trebouxia.
DISTRIBUTION: Queensland, New South Wales, Victoria and South Australia. 3 species reported (7).

CANDELARIELLA MULL. ARG.: *Bull. Herb. Boissier* 2 append 1, 11 (1894)

MORPHOLOGY: Thallus crustose, uniform or granular or lobed at the margins, pale or deep yellow from pulvinic acid derivatives; attached to the substrate by hyphae; upper surface paraplectenchymatous. *Ascocarp* an apothecium, sessile, circular, thalloid margin; hypothecium pale, with algae below; paraphyses usually unbranched; asci unitunicate, thick walled, clavate, I+ blue, tholus I+ blue, 8-many spored; ascospores simple or 1 septate, hyaline, thin walled. *Pycnidia* immersed; fulcrum exobasidial; pycnospores bacilliform, somewhat constricted at the centre.
CHEMISTRY: (4 species) calycin (4), pulvinic dilactone (4), stictaurin (3).
PHYCOBIONT: Trebouxia.
DISTRIBUTION: South Australia and Victoria. 2 species reported (40).

CATILLARIA MASS. em. TH. FR.: *Lich. Scand.* 563 (1874)

MORPHOLOGY: Thallus crustose, often granular; ecorticate; attached by hyphal rhizoids. *Ascocarp* an apothecium, immersed to sessile; disk round, pale to dark, hypothecium pale or dark; margin and hypothecium without algae; paraphyses unbranched, free, sometimes capitate; ascus clavate-cylindrical, unitunicate, tholus I+ blue, 8-spored; ascospores 1 septate, thin walled, hyaline. *Pycnidia* immersed; fulcrum exobasidial; pycnospores oblong to pyriform or curved.
CHEMISTRY: (3 species) usnic acid (2), atranorin (1).
PHYCOBIONT: ?Trebouxia, ?Myrmecia.
DISTRIBUTION: Widespread, especially on smooth barked trees. 31 species reported (300).

CATINARIA VAIN.: *Acta Soc. Fauna Flora fenn.* 53(1), 143 (1922)

MORPHOLOGY: Thallus crustose, uniform, ecorticate, attached by medullary or prothallial hyphae. *Ascocarp* an apothecium-like pseudothecium, more or less immersed; circular or somewhat irregular; initially with a proper exciple which later disappears; hypothecium black; paraphyses often reticulate, asci clavate, bitunicate, I−, 8 spored; ascospores oblong or ellipsoid, 1 septate, hyaline, slightly constricted at the septum, walls somewhat thickened. *Pycnidia* not known.
CHEMISTRY: No data.
PHYCOBIONT: Chlorophyceae.
DISTRIBUTION: Victoria. 1 species reported (?).

CETRARIA ACH.: *Meth. Lich.* 292 (1803)

MORPHOLOGY: *Thallus* foliose to fruticose, more or less ascending from the substrate, often markedly flattened and strap-like, sometimes cylindrical; all surfaces corticate; lower surface often pseudocyphellate, without rhizines; margins sometimes spiny; free, partly buried or attached by a holdfast. *Ascocarp* an apothecium, marginal on lobes, round; margin thalline; hypothecium pale; paraphyses unbranched or rarely branched; asci cylindrical, unitunicate, I+ blue, tholus I+ blue, 8 spored; ascospores ellipsoidal to spherical, simple, hyaline, thin or thick walled. *Pycnidia* sessile; fulcrum endobasidial, short; pycnospores ellipsoidal to cylindrical, or bacilliform.
CHEMISTRY: (58 species) protolichesterinic acid (15), usnic acid (15), atranorin (7), vulpinic acid (6), alectoronic acid (5), caperatic acid (5), fumarprotocetraric acid (5), physodic acid (4), pinastric acid (4), norrangiformic acid (3), rangiformic acid (3), stictic acid (3), α-collatolic acid (2), endocrocin (2), friedelin (2), lichesterinic acid (2), olivetoric acid (2), *allo*-protolochesterinic acid (2), gyrophoric acid (1), hiascic acid (1), microphyllinic acid (1), nephrosteranic acid (1), norstictic acid (1).
PHYCOBIONT: *Trebouxia.*
DISTRIBUTION: New South Wales, Victoria and Tasmania, in alpine areas. 2 species reported (45).

CETRELIA W. CULB. et C. CULB.: *Contr. U.S. Natn. Herb.* 34 (1968)

MORPHOLOGY: *Thallus* foliose, lobes to 2.5 cm broad, pale above; upper surface prosoplectenchymatously corticate, with punctiform pseudocyphellae, isidate or sorediate; lower surface black, shiny, corticate, rhizinate; without marginal cilia. *Ascocarp* an apothecium, round, stipitate, sub-marginal, usually perforate; hypothecium hyaline; paraphyses unbranched or little branched; asci cylindrical, unitunicate, thick walled, I+ blue, tholus I+ blue, 8 spored; ascospores ellipsoid, simple, hyaline. *Pycnidia* marginal; fulcrum ?endobasidial; pycnospores bacilliform.
CHEMISTRY: (14 species) atranorin (14), imbricaric acid (4) alectoronic acid (3), α-collatolic acid (3), olivetoric acid (3), anziaic acid (2), perlatolic acid (2), microphyllinic acid (1).
PHYCOBIONT: *Trebouxia.*
DISTRIBUTION: Queensland, on bark. 1 species reported (14).

CHAENOTHECA TH. FR.: *Lich. Scand.* 563 (1874)

MORPHOLOGY: *Thallus* crustose to rarely squamulose or endophloic, ecorticate, attached by hyphae. *Ascocarp* an apothecium on a stipe; disk round, often deeply concave; margin without algae, dark; paraphyses unbranched; asci cylindrical, unitunicate, thin walled, 8 spored; ascospores simple, globose, dark coloured; asci and paraphyses disintegrate leaving the spores lying free in a mazaedium. *Pycnidia* punctiform; fulcrum exobasidial; pycnospores bacilliform.
CHEMISTRY: (1 species) vulpinic acid (1).
PHYCOBIONT: *Stichococcus, Trentepohlia.*
DISTRIBUTION: Victoria and Tasmania. 3 species reported (23).

CHIODECTON ACH.: *Syn. Meth. Lich.* 108 (1814)

MORPHOLOGY: *Thallus* crustose, ecorticate, uniform, undifferentiated; attached by prothallial hyphae. *Ascocarp* a pseudothecium, immersed or sessile in groups in stroma;

hymenium open, ovoid to stellate; margin well developed, black, without algae, or rudimentary; hypothecium black; paraphyses reticulate; asci bitunicate, clavate, I−, 8 spored; ascospores fusiform to acicular, transversely many septate, hyaline, walls thin, locules cylindrical. *Pycnidia* immersed; fulcrum exobasidial; pycnospores cylindrical, oblong to ellipsoid, straight or curved.
CHEMISTRY: (4 species) atranorin (1), erythrin (1), gyrophoric acid (1), norstictic acid (1), psoromic acid (1), salazinic acid (1).
PHYCOBIONT: *Trentepohlia.*
DISTRIBUTION: Queensland, New South Wales, Victoria and Tasmania. 19 species reported (175).

CHONDROPSIS NYL. in CROMBIE: *J. Linn. Soc. Bot.* 17: 397 (1879)

MORPHOLOGY: *Thallus* foliose, repeatedly dichotomously branched; devoid of attachment to the substrate; upper cortex prosoplectenchymatous, opaque when dry, transparent when wet; lower cortex of interwoven hyphae; the whole rolling into a ball when dry with the lower cortex outermost. *Ascocarp* an apothecium; rare, sessile, margin thalline; disk round, concave to flat, light brown; hypothecium pale; paraphyses unbranched; asci cylindrical, unitunicate, tholus I+ blue, 8 spored; ascospores simple, hyaline. *Pycnidia* not known.
CHEMISTRY: (monospecific genus) fumarprotocetraric acid, succinprotocetraric acid, usnic acid.
PHYCOBIONT: *Trebouxia.*
DISTRIBUTION: On soil, in semi-arid southern Australia, and the Snowy mountains. 1 species (1).

CHRYSOTHRIX MONT.: *Annls Sci. nat. (Bot.)* Ser. 3, 18, 312 (1852)

MORPHOLOGY: *Thallus* a tangled weft of hyphae enclosing soredium-like clusters of algae, forming a crust or small polsters; attached to the substrate by hyphae. *Ascocarp* very rarely seen, immersed in the thallus or sessile, disk round; margin proper; hypothecium hyaline; paraphyses disappearing; asci 8-spored; ascospores transversely septate, hyaline. *Pycnidia* not known.
CHEMISTRY: (3 species) calycin (3), vulpinic acid (2), leprapinic acid (1), leprapinic acid methyl ether (1), pinastric acid (1).
PHYCOBIONT: *Coccoid green algae.*
DISTRIBUTION: On wood and bark throughout the region. 1 species reported (3).

CLADIA NYL.: *Bull. Soc. Linn. Normandie ser 2,* 6: 167 (1870)

MORPHOLOGY: *Primary thallus* absent; pseudopodetia much or little branched; esquamulose, esorediose, non-isidiose, often reticulately fenestrate; cortex of longitudinal hyphae, medulla variable. *Ascocarp* an apothecium; apothecia often in cymose stacks, terminal on pseudopodetia; disk round; margin prominent, dark, without algae; hypothecium pale or dark; paraphyses unbranched; asci clavate, unitunicate, tholus I+ blue, 8 spored; ascospores simple, hyaline. *Pycnidia* sessile or substipitate; fulcrum exobasidial; pycnospores curved.
CHEMISTRY: (7 species) ursolic acid (7), usnic acid (4), protocetraric acid (4), protolichesterinic acid (4), atranorin (3), divaricatic acid (2), fumarprotocetraric acid (2), norstictic acid (1), barbatic acid (1), 4-0-demethylbarbatic acid (1), gyrophoric acid (1), homosekikaic acid (1).

PHYCOBIONT: *Trebouxia.*
DISTRIBUTION: Widespread on soils and rotting wood, across southern Australia (including southern Queensland). 7 species reported (7).

CLADINA NYL. in OLIV.: *Expos Lich. Ouest France* 1: 41 (1897)

MORPHOLOGY: *Primary thallus* crustose, ecorticate, short lived; *Podetia* much branched, hollow, cylindrical ecorticate, without squamules. *Ascocarp* an apothecium, terminal on podetia, cymose, minute, pale or brown, margin proper; hypothecium pale; paraphyses unbranched or little branched; asci clavate-cylindrical, unitunicate, tholus I+ blue, 8 spored; ascospores simple, hyaline. *Pycnidia* terminal on podetia; fulcrum exobasidial; pycnospores filiform, curved or straight.
CHEMISTRY: (6 species) usnic acid (4), fumarprotocetraric acid (3), perlatolic acid (2), ursolic acid (2), atranorin (1).
PHYCOBIONT: *Trebouxia.*
DISTRIBUTION: On soil in cooler, commonly alpine or sub-alpine, areas in New South Wales, Victoria and Tasmania. 2 species reported (10).

CLADONIA HILL.: *Hist. pl.* 91 (1773)

MORPHOLOGY: *Primary thallus* squamulose to foliose, corticate, upper cortex of thick walled hyphae perpendicular to the surface, attached to the substrate by hyphal rhizines. *Podetia* common, laminal or rarely marginal, simple or branched, often cup-shaped, appearing fruticose, naked or covered in squamules or soredia, corticate or ecorticate, hollow, axils open or perforate. *Ascocarp* an apothecium, raised on podetia, terminal or on margins of cups, rarely sessile or squamules, disk markedly convex, pale brown, dark brown or red; margin proper, not prominent; hypothecium colourless; paraphyses unbranched or little branched; Asci clavate-cylindrical, unitunicate, tholus I+ blue, 8 spored; ascospores simple, hyaline, thin walled. *Pycnidia* terminal on podetia or sessile or stipitate on squamules or on cups; fulcrum exobasidial; pycnospores filiform, straight or curved.
CHEMISTRY: (94 species) usnic acid (35), fumarprotocetraric acid (27), atranorin (14), squamatic acid (13), barbatic acid (12), thamnolic acid (9), bellidiflorin (6), baeomycesic acid (4), didymic acid (4), homosekikaic acid (4), ursolic acid (4), psoromic acid (4), destrictinic acid (3), constictic acid (2), merochlorophaeic acid (2), norstictic acid (2), perlatolic acid (2), porphyrilic acid (2), protolichesterinic acid (2), stictic acid (2), succinprotocetraric acid (2), zeorin (2) bourgeanic acid (1), caperatic acid (1), cryptochlorophaeic acid (1), diacetylgraciliformin (1), decarboxythamnolic acid (1), graciliformin (1), grayanic acid (1), imbricaric acid (1), lichesterinic acid (1), 4-0-methyl-cryptochlorophaeic acid (1), monoactylgraciliformin (1), rhodocladonic acid (1), skyrin (1), strepsilin (1).
PHYCOBIONT: *Trebouxia.*
DISTRIBUTION: In all states, especially common in southern regions, on soil and rotting logs. 50 species reported (300)

CLATHROPORINA MULL. ARG.: *Flora, Jena* 65, 517 (1882)

MORPHOLOGY: *Thallus* crustose, epi- or endophloic, uniform, ecorticate. *Ascocarp* a perithecium, sessile to somewhat immersed, not aggregated, ostiole erect; paraphyses unbranched; asci thin walled, unitunicate, clavate, I−, 8 spored; ascospores ellipsoid to fusiform, muriform, hyaline, walls thin, of even thickness. *Pycnidia* immersed; fulcrum exobasidial; pycnospores oblong to oblong-cylindrical.

CHEMISTRY: No data.
PHYCOBIONT: *Trentepohlia.*
DISTRIBUTION: On trees, in Queensland and Tasmania. 9 species reported (60).

COCCOCARPIA PERS. in GAUD.: *Voyage Uranie et Physicienne* 206 (1826)

MORPHOLOGY: *Thallus* squamulose to foliose, often much lobed, attached by a rhizoidal tomentum, upper surface with a cortex composed of longitudinally interwoven hyphae, algal layer immediately below the upper cortex, medulla grading into lower cortex. *Ascocarp* an apothecium, sessile; proper margin barely visible; devoid of algae; hypothecium pale or dark; paraphyses unbranched; asci unitunicate, clavate, tholus I+ blue, 8 spored; ascospores simple, globose to fusiform, thin walled, hyaline. *Pycnidia* immersed in thalline warts; fulcrum endobasidial, much septate; pycnospores bacilliform.
CHEMISTRY: (4 species) no lichen substances.
PHYCOBIONT: *Scytonema.*
DISTRIBUTION: On trees and rocks, in Queensland, New South Wales and Victoria. 5 species reported (25).

COCCOTREMA MULL. ARG.: *Miss. Sci. Cap Horn* vol. 5: 171 (1889)

MORPHOLOGY: *Thallus* crustose, uniform, attached to the substrate by medullary hyphae. *Ascocarp* an apothecium, immersed in fruiting warts, without a proper margin; hymenium and receptacle clearly but narrowly perforate; paraphyses indistinct, unbranched, branched or reticulate; hypothecium hyaline; asci clavate, unitunicate (but of several distinct layers), thick walled, I+ blue, tholus I+; ascospores simple, hyaline, not greatly thickened. *Pycnidia* not known.
CHEMISTRY: (3 species) constictic acid (2), stictic acid (2), norstictic acid (1).
PHYCOBIONT: *Trebouxia*-like or *Trentepohlia*-like.
DISTRIBUTION: On bark, in Tasmania. 1 species reported (6).

COELOCAULON LINK: *Grund. der Krauterkunde* vol. III, 165 (1833)

MORPHOLOGY: *Thallus* ascending, fruticose, fragile, cylindrical in cross-section, hollow attached by sparse rhizines, corticate or dissolving into soredia. *Ascocarp* not known. *Pycnidia* not known.
CHEMISTRY: (2 species) olivetoric acid (1), protolichesterinic acid (1).
PHYCOBIONT: *Trebouxia.*
DISTRIBUTION: On rocks, in South Australia and Western Australia. 1 species reported (7).

COENOGONIUM EHRENB. ex NEES.: *Horae Physic. Berol.* 120 (1820)

MORPHOLOGY: *Thallus* a spongy filamentous mass forming a mound or disk, either shelf-like or pendant from the substrate; hyphae entangling the phycobiont. *Ascocarp* an apothecium, laminal or marginal, short stalked; margin proper, pseudoparenchymatous; paraphyses loose, unbranched; asci unitunicate, thin walled, I+ blue, 8 spored, ascospores simple or 1 septate, hyaline, uniformly thin walled. *Pycnidia* globose; fulcrum exobasidial, bushy; pycnospores fusiform, straight.
CHEMISTRY: No data.

PHYCOBIONT: *Trebouxia.*
DISTRIBUTION: Queensland, Victoria and Tasmania. 9 species reported (15).

COLLEMA WEB. in WIGG.: *Prim. Flor. Holsat.* 89 (1780)

MORPHOLOGY: *Thallus* foliose, sometimes almost crustose, gelatinous, ecorticate, comprising a mycelial weft through an algal gelatinous matrix, directly attached to the substrate. *Ascocarp* an apothecium, immersed to sessile; disk round; margin thalline; hypothecium pale; paraphyses unbranched; asci clavate, unitunicate, I+ blue, 8 spored; ascospores transversely 1-many septate or muriform, hyaline, thin walled. *Pycnidia* immersed in the thallus or in warts; fulcrum endobasidial, short celled; pycnospores bacilliform.
CHEMISTRY: Without characteristic lichen substances.
PHYCOBIONT: *Nostoc.*
DISTRIBUTION: On soil, rocks and trees, throughout Australia. 12 species reported (80).

CONIOCYBE ACH.: *Kgl. Vetensk-Akad. Nya Handl.* 37, 286 (1815)

MORPHOLOGY: *Thallus* crustose, powdery or disappearing, ecorticate. *Ascocarp* an apothecium, stipitate on long fine stalks; disk initially flat, becoming globose as the proper exciple is displaced; paraphyses thread-like; asci cylindrical, thin walled, unitunicate, 8 spored; ascospores globose, simple, yellowish or hyaline; asci and paraphyses disintegrating leaving the spores free in a powdery mass or mazaedium. *Pycnidia* protruding from the thallus; fulcrum exobasidial; pycnospores oblong-ellipsoid.
CHEMISTRY: No data.
PHYCOBIONT: *Stichococcus.*
DISTRIBUTION: Victoria. 4 species reported (20).

CONOTREMOPSIS VEZDA: *Folia Geobot. Phytotax. Praha* 12, 313 (1977)

MORPHOLOGY: *Thallus* filamentous, woolly. *Ascocarp* an apothecium, narrowly cylindrical, closed at first then opening, disk round; thalline margin disappearing leaving a thick proper margin; hypothecium dark; paraphyses unbranched; asci cylindrical, unitunicate, I−; ascospores long and filamentous, transversely many (up to 100) septate, hyaline, thin walled. *Pycnidia* not known.
CHEMISTRY: No data.
PHYCOBIONT: *Trentepohlia.*
DISTRIBUTION: Known only from type collected in Tasmania. 1 species reported (1).

CROCYNIA (ACH.) MASS.: *Atti I.R. Ist. Veneto Ser 3,* 5, 251 (1860)

MORPHOLOGY: *Thallus* lobed, of interwoven hyphal filaments, almost membranous, with a distinct prothallus, without rhizines, ecorticate; the phycobiont loosely entangled in a hyphal weft. *Ascocarp* an apothecium, laminal, shortly stipitate; disk circular; margin proper, well developed; hypothecium hyaline; paraphyses obscure; asci cylindrical, unitunicate, tholus I+ blue, 8 spored; ascospores simple, hyaline, thin walled. *Pycnidia* not known.
CHEMISTRY: (1 species) usnic acid, zeorin.
PHYCOBIONT: *?Trebouxia, ?Chlorella, Stichoccus.*
DISTRIBUTION: On bark, in Victoria and Tasmania. 3 species reported (?)

CYPHELIUM ACH.: *K. svensk Vetensk Akad, Handl.* 261 (1815)

MORPHOLOGY: Thallus crustose, powdery, uniform, ecorticate. *Ascocarp* an apothecium, almost immersed in thalline tissue or sessile; disk at first almost closed, later open; margin black, proper, with or without thalline rim; paraphyses thread-like; asci cylindrical, thin walled, unitunicate 8-spored; ascospores 1-septate (rarely simple), dark; asci and paraphyses disintegrating leaving the spores free in a powdery mass or mazaedium. *Pycnidia* small, fulcrum endobasidial (?); pycnospores bacilliform.
CHEMISTRY: (10 species) rhizocarpic acid (5), atranorin (2), chloroatranorin (2), usnic acid (2), epanorin (1), norstictic acid (1), vulpinic acid (1).
PHYCOBIONT: ?Stichococcus.
DISTRIBUTION: Queensland, Victoria and Tasmania, on wood and bark. 5 species reported (30).

CYSTOCOLEUS THW.: *Ann. Nat. Hist. Ser 2,* 3, 241 (1849)

MORPHOLOGY: Thallus erect, of radiating filaments comprised of algal cells entwined in fungal hyphal cylinders. *Ascocarp* unknown. *Pycnidia* unknown.
CHEMISTRY: No data.
PHYCOBIONT: Trentepohlia.
DISTRIBUTION: Tasmania. 1 species reported (1 or 2).

DENDRISCOCAULON NYL.: *Flora, Jena* 68, 299 (1885)

MORPHOLOGY: Thallus fruticose, erect, cylindrical or somewhat flattened, often much branched or arborescent; cortex unevenly developed, sometimes with cyphella-like pores; sometimes bearing flattened lobes which can be referred to species of *Sicta* or *Lobaria*, and which may be interpreted as cephalodia containing green algae; attached by a basal holdfast. *Ascocarp* not known. *Pycnidia* not known.
CHEMISTRY: (1 species) no secondary lichen substances.
PHYCOBIONT: Nostoc.
DISTRIBUTION: Victoria and Tasmania, in wet areas. 1 species reported (?6).

DERMATOCARPON ESCHW.: *Syst. Lich.* 21 (1824)
(including *Catapyrenium* Fee.)

MORPHOLOGY: Thallus squamulose or foliose, apressed or somewhat ascending; upper or upper and lower surfaces paraplectenchymatously corticate; attached by rhizoids or prothallial hyphae. *Ascocarp* a perithecium, immersed in the thallus; hymenium without algae; paraphyses reticulately branched, soon gelatinizing; asci unitunicate, thick walled, I−, 8 spored; ascospores simple, hyaline. *Pycnidia* immersed; fulcrum endobasidial; pycnospores bacilliform.
CHEMISTRY: (5 species) no secondary lichen substances.
PHYCOBIONT: Myrmecia, Hyalococcus.
DISTRIBUTION: On soil and rock, in Queensland, New South Wales, Victoria, South Australia and Western Australia. 4 species reported (80).

DERMATINA ALMQ.: *Kgl. Svensk Vetensk-Akad. Handl.* 17(6), 8 (1880)

MORPHOLOGY: Thallus crustose, uniform, ecorticate. *Ascocarp* a perithecium-like pseudothecium, opening by a pore or irregular slit, often several fused together immersed in a stroma, black; paraphyses reticulate, often disintegrating; asci oval or clavate, bitunicate, thick walled, I−, 8 spored; ascospores ellipsoid, hyaline or brown, muriform, thin walled. *Pycnidia* immersed; fulcrum exobasidial; pycnospores oblong, cylindrical.
CHEMISTRY: No data.
PHYCOBIONT: Palmella-like.
DISTRIBUTION: Victoria. 1 species reported (30).

DICTYONEMA C.AG. ex KUNTH.: *Syn. Pl. Aequinoct.* vol. 1, 1 (1822)

MORPHOLOGY: Thallus initially of tangled hyphae which form a crust and eventually semi-circular brackets standing out from the substrate; the upper surface becomes shaggy with concentric markings; lower surface eventually covered by a hymenium; composed of thin hyaline articulate hyphae which are irregularly branched or at times dichotomous. *Hymenium* confined to the lower surface bearing an irregular layer of basidia; basidia bear four simple hyaline or yellowish-brown spores.
CHEMISTRY: (4 species) no secondary lichen substances found.
PHYCOBIONT: Scytonema or *Chroococcus.*
DISTRIBUTION: In rainforests in Queensland. 2 species reported (5).

DIMERELLA TREVIS.: *Rc. Ist. Lomb. Sci. Lett. ser. 2,* 13, 65 (1880)

MORPHOLOGY: Thallus crustose, effuse, uniform, ecorticate. *Ascocarp* an apothecium, sessile; disk round; margin proper, yellow or orange; hypothecium, hyaline; paraphyses unbranched; asci cylindrical to clavate, unitunicate, thin walled, I+ blue, 8 spored; ascospores oblong to fusiform, 1 septate, hyaline. *Pycnidia* immersed or sessile; fulcrum endobasidial; pycnospores bacilliform, 1 septate.
CHEMISTRY: No data.
PHYCOBIONT: Trentepohlia.
DISTRIBUTION: On trees, in Queensland, New South Wales, Victoria and Tasmania. 4 species reported (25).

DIPLOGRAMMA MULL. ARG.: *Nuov. Giorn. Bot. Ital.* 23, 399 (1891)

MORPHOLOGY: Thallus endophloic, ecorticate, uniform. *Ascocarp* an apothecium, sessile, elongate, with twin longitudinal hymenia; margin carbonaceous; hymenia black, narrow; paraphyses reticulately branched; asci 8 spored; ascospores transversely septate, hyaline. *Pycnidia* not known.
CHEMISTRY: No data.
PHYCOBIONT: Palmella-like.
DISTRIBUTION: On bark, in Queensland. 1 species reported (1).

DIPLOICIA MASS.: *Ric. Lich. Crost.* 86 (1852)

MORPHOLOGY: Thallus squamulose; corticate; attached by medullary hyphae. *Ascocarp* an

apothecium, sessile; disk round, margin proper, black; hypothecium dark; paraphyses unbranched; asci unitunicate, I+ blue, tholus I+ blue, 8 spored; ascospores transversely 1-3 septate, brown. *Pycnidia* immersed or in thalline warts; fulcrum endobasidial; pycnospores bacilliform.
CHEMISTRY: (1 species) atranorin, chloroatranorin, diploicin.
PHYCOBIONT: Trebouxia.
DISTRIBUTION: On soil and rocks, in New South Wales, South Australia, Victoria and Western Australia. 2 species reported (4).

DIPLOSCHISTES NORM.: *Nytt. Mag. Naturv.* 7, 213 (1853)

MORPHOLOGY: Thallus crustose, often areolate, without lobes, ecorticate or with a cortex of tangled hyphae; attached by medullary hyphae. *Ascocarp* an apothecium, immersed to sessile; disk closed or open; proper exciple well developed; thalline exciple often well developed; paraphyses unbranched or branched near the apices; asci clavate, unitunicate, thick walled, I+, 2-8 spored; spores muriform, brown, thin walled. *Pycnidia* immersed; fulcrum endobasidial (?); pycnospores bacilliform.
CHEMISTRY: (22 species) lecanoric acid (14), diploschistesic acid (9), atranorin (6), connorstictic acid (2), gyrophoric acid (2), norstictic acid (2), ursolic acid (2), ethyl orsellinate (1), fumarprotocetraric acid (1), orsellinic acid (1), zeorin (1).
PHYCOBIONT: Coccoid Chlorophyceae.
DISTRIBUTION: Across the southern half of the continent and Tasmania, on soil and rock. 6 species reported (35).

DIRINARIA TUCK.: *Proc. Amer. Acad. Arts & Sci.* 12, 166 (1877)

MORPHOLOGY: Thallus foliose, closely apressed to the substrate, attached by rhizines, upper cortex paraplectenchymatous. *Ascocarp* an apothecium, sessile; disk concave to convex, round, thalline margin; hypothecium dark; paraphyses unbranched; asci clavate, unitunicate, I+ blue, tholus I+ blue, 8 spored; ascospores 1 septate, brown, thick walled. *Pycnidia* immersed; fulcrum endobasidial; pycnospores short, cylindrical.
CHEMISTRY: (26 species) atranorin (26), divaricatic acid (19), sekikaic acid (6), triterpenoids (5), zeorin (2), chloroatranorin (1), lecanoric acid (1), ramalinolic acid (1).
PHYCOBIONT: Trebouxia.
DISTRIBUTION: On rocks and trees, especially in the northern half of the continent. 8 species reported (26).

DIRINASTRUM MULL. ARG.: *Bull. Herb. Boissier* 1, 55 (1893)

MORPHOLOGY: Thallus crustose, uniform; cortex of anticlinal hyphae; attached by medullary hyphae. *Ascocarp* an apothecium-like pseudothecium, sessile to immersed, round or elongate; proper and thalline margins well developed; hypothecium black; paraphyses unbranched; asci bitunicate, I−, 8 spored; ascospores transversely 3-7 septate, brown. *Pycnidia* immersed; fulcrum exobasidial; pycnospores bacilliform; curved.
CHEMISTRY: (1 species), erythrin.
PHYCOBIONT: Trentepohlia.
DISTRIBUTION: Coastal rocks, in Victoria. 1 species reported (1).

ECHINOPLACA FEE: *Essai Crypt.* 50 (1824)

MORPHOLOGY: Thallus crustose, uniform, often with white hairs, with a developed cortex. *Ascocarp* an apothecium, immersed to sessile, round to oval, exciple poorly developed or absent; paraphyses reticulate; asci thick walled, unitunicate, 1-8 spored; ascospores transversely several septate or muriform, hyaline, thin walled. *Pycnidia* not known.
CHEMISTRY: No data.
PHYCOBIONT: Trebouxia-like.
DISTRIBUTION: On leaves, in Queensland. 3 species reported (9).

ENDOCARPON HEDW.: *Descr. et adum. Musc.* 56 (1789)

MORPHOLOGY: Thallus squamulose to rarely foliose; upper surface paraplectenchymatous; attached by rhizoids. *Ascocarp* a perithecium, immersed, ostiole erect; with hymenial algae; paraphyses gelatinizing; asci globose, unitunicate, thick walled, 1-8 spored; ascospores muriform, hyaline to brown, thin walled. *Pycnidia* immersed; fulcrum exobasidial; pycnospores cylindrical, straight.
CHEMISTRY: No data.
PHYCOBIONT: Pleurococcus, Stichococcus.
DISTRIBUTION: On soil, in New South Wales, Victoria, South Australia and Western Australia. 4 species reported (30).

ENTERODICTYON MULL. ARG.: *J. Linn. Soc. Bot.* 29, 230 (1892)

MORPHOLOGY: Thallus crustose, uniform, ecorticate, attached by medullary hyphae. *Ascocarp* an apothecium, sessile in stroma; disk open, irregular; proper exciple brown, well or poorly developed; paraphyses unbranched; asci bitunicate, I−, 8 spored; spores muriform, hyaline, thin walled. *Pycnidia* not known.
CHEMISTRY: No data.
PHYCOBIONT: Trentepohlia.
DISTRIBUTION: On bark, in New South Wales. 1 species reported (3).

ENTEROGRAPHA FEE:: *Essai Crypt.* 32 (1824)

MORPHOLOGY: Thallus crustose, uniform, of densely interwoven hyphae, ecorticate; attached by medullary or prothallial hyphae. *Ascocarp* a pseudothecium, immersed in stroma, single or grouped; exciple absent or poorly developed; hypothecium pale; paraphyses reticulate; asci bitunicate, clavate, I−, 8 spored; ascospores transversely many septate, hyaline, walls thin, locules cylindrical. *Pycnidia* not known.
CHEMISTRY: (1 species) confluentic acid.
PHYCOBIONT: Phycopeltis or *Trentepohlia.*
DISTRIBUTION: On bark, in Victoria. 1 species reported (30).

EPHEBE FR.: *Syst. orb.* 256 (1825)

MORPHOLOGY: Thallus filamentous, ramifying in the gelatinous sheath of the phycobiont, appearing minutely fruticose. *Ascocarp* an apothecium, immersed in thickenings in the filaments, almost globose, punctiform; disk almost closed; proper exciple thin; hypothecium

hyaline, paraphyses absent; asci clavate, unitunicate, I+ blue, 8 spored; ascospores simple or 1 septate, hyaline, walls thin. *Pycnidia* immersed in globose swellings; fulcrum exobasidial, filiform; pycnospores bacilliform.
CHEMISTRY: (1 species) no secondary lichen substances.
PHYCOBIONT: *Stigonema.*
DISTRIBUTION: On rocks, in New South Wales, Victoria, Tasmania, South Australia and Western Australia. 3 species reported (12).

ERIODERMA FEE: *Essai Crypt.* 145 (1824)

MORPHOLOGY: *Thallus* foliose, lobed, corticate; upper cortex pseudoparenchymatous, shaggy; lower surface ecorticate, tomentose, sometimes veined; attached by laminal or marginal rhizines. *Ascocarp* an apothecium, marginal or laminal, sessile to stipitate; proper exciple well developed; disk round; hypothecium pale; asci unitunicate, clavate, I+ blue, 8 spored; ascospores simple, hyaline. *Pycnidia* marginal, sessile; fulcrum endobasidial; pycnospores bacilliform.
CHEMISTRY: (4 species) unidentified depsidone like pannarin.
PHYCOBIONT: *Scytonema.*
DISTRIBUTION: On bark, in Queensland. 1 species reported (16).

EVERNIA ACH.: *Lich. Univ.* 84 (1810)

MORPHOLOGY: *Thallus* fruticose to apparently foliose, upright or pendulous, dorsiventral; both surfaces corticate, cortex of anticlinal hyphae; without rhizines, attached by a basal disk. *Ascocarp* an apothecium, lateral or terminal, stipitate, concave, thalline margin; hypothecium hyaline, paraphyses unbranched; asci clavate, unitunicate, tholus I+ blue, 8 spored; ascospores simple, hyaline, thin walled. *Pycnidia* marginal, immersed; fulcrum endobasidial; pycnospores acicular.
CHEMISTRY: (4 species) divaricatic acid (4), usnic acid (2), atranorin (1), chloroatranorin (1), evernic acid (1).
PHYCOBIONT: *Trebouxia.*
DISTRIBUTION: In alpine/sub-alpine areas of New South Wales. 1 species reported (6).

FISTULARIELLA BOWLER & RUNDEL: *Mycotaxon* 6, 195 (1977)

MORPHOLOGY: *Thallus* fruticose, erect, hollow and inflated, little branched, corticate, fenestrate; attached by a basal disk. *Ascocarp* an apothecium, usually appearing terminal, concave; well developed thalloid exciple; disk round; hypothecium pale; paraphyses unbranched; asci clavate, unitunicate, tholus I+ blue, 8 spored; ascospores 1 septate, thin walled, hyaline. *Pycnidia* immersed in the thallus; fulcrum exobasidial, little branched; pycnospores bacilliform, straight.
CHEMISTRY: (5 species) usnic acid (4), sekikaic acid (3), divaricatic acid (2), norstictic acid (2), evernic acid (1), obtusatic acid (1).
PHYCOBIONT: *Trebouxia.*
DISTRIBUTION: On twigs, wood or soil and rock, throughout Australia except in very dry inland areas. 8 species reported (27).

FULGENSIA MASS. ex DE NOT. in MASS.: *Gen. Lich.* 10 (1854b)

MORPHOLOGY: Thallus crustose, granular, lobed at the margins. *Ascocarp* an apothecium, sessile; thalloid exciple developed; disk round, orange, hypothecium pale; paraphyses unbranched, thickened at the tips; asci unitunicate, I+ blue, tholus I+ blue, 8 spored; ascospores simple, hyaline. *Pycnidia* immersed; fulcrum endobasidial; pycnospores bacilliform.
CHEMISTRY: (9 species) parietin (9), emodin (8), fragilin (4), teloschistin (3), 2-chloroemodin (2), fallacinal (2), parietinic acid (2).
PHYCOBIONT: Trebouxia.
DISTRIBUTION: On dry calcareous substrates, in the southern half of the continent and Tasmania. 1 species reported (10).

GLYPHIS ACH.: *Syn. Lich.* 106 (1814)

MORPHOLOGY: Thallus crustose, uniform; ecorticate or with a poorly developed cortex of periclinal hyphae; attached by medullary or prothallial hyphae. *Ascocarp* an apothecium, elongate or irregular; immersed in stroma; well developed proper exciple; hypothecium pale; hymenium gelatinous; paraphyses unbranched; asci oblong, unitunicate, I−, 8 spored; ascospores transversely 3-11 septate, walls unequally thickened to produce lenticular locules, hyaline, I+ violet. *Pycnidia* not know.
CHEMISTRY: No data.
PHYCOBIONT: Trentepohlia.
DISTRIBUTION: On bark, in Queensland. 3 species reported (30).

GRAPHINA MULL. ARG.: *Flora, Jena* 63, 22 (1880)

MORPHOLOGY: Thallus crustose, uniform, epi- or endophloic; ecorticate or with a cortex of interwoven hyphae; attached by medullary or prothallial hyphae. *Ascocarp* an apothecium, elongate, often branched; adnate or immersed; disk narrow; proper exciple usually well developed; hypothecium pale to black; paraphyses unbranched; asci clavate, unitunicate, I−, usually only 1-3 spored; ascospores oblong-ellipsoid to elongate-ellipsoid, large, muriform, hyaline, thin walled, I+ violet. *Pycnidia* rare, immersed; fulcrum exobasidial; pycnospores oblong-cylindrical to bacilliform.
CHEMISTRY: (19 species) norstictic acid (6), stictic acid (6), atranorin (3), protocetraric acid (3), lichexanthone (2), salazinic acid (2), zeorin (1).
PHYCOBIONT: Trentepohlia.
DISTRIBUTION: On wood and bark, throughout the region. 18 species reported (270).

GRAPHIS ADANS.: *Fam. Pl.* 11 (1763)

MORPHOLOGY: Thallus crustose, uniform, epi- or endophloic; ecorticate or with a cortex of interwoven hyphae; attached by medullary or prothallial hyphae. *Ascocarp* an apothecium, immersed to sessile, elongate, often branched, disk narrow; proper exciple often black; thalloid exciple sometimes present; hypothecium pale to black; paraphyses unbranched; asci clavate, unitunicate, I−, thick walled, 8 spored; ascospores transversely 3-many septate, hyaline, walls unequally thickened resulting in lenticular locules, I+ violet. *Pycnidia* rare, immersed; fulcrum exobasidial; pycnospores oblong cylindrical to bacilliform.
CHEMISTRY: (16 species) norstic acid (8), stictic acid (4), protocetraric acid (2), atranorin (1), lecanoric acid (1), salazinic acid (1).

PHYCOBIONT: *Trentepohlia.*
DISTRIBUTION: On wood and bark, throughout the region. 41 species reported (300).

GYALECTA ACH.: *Kgl. Vetensk Akad. Nya Handl.* 29, 228 (1808)

MORPHOLOGY: *Thallus* crustose, uniform; attached to the substrate by prothallial hyphae; ecorticate. *Ascocarp* an apothecium, circular; immersed to sessile; proper exciple pale, waxy; thalloid exciple often present; hypothecium pale, soft; paraphyses unbranched; asci clavate, unitunicate, thin walled, I+ blue, without a tholus, 8 spored; ascospores transversely 3-11 septate or muriform, hyaline, thin walled. *Pycnidia* immersed; fulcrum exobasidial; pycnospores linear or bacilliform.
CHEMISTRY: No data.
PHYCOBIONT: *Trentepohlia, Gloeocystis.*
DISTRIBUTION: In Victoria and Tasmania. 1 species reported (34).

GYALECTIDIUM MULL. ARG.: *Flora, Jena* 64, 100 (1881)

MORPHOLOGY: *Thallus* crustose, effuse or with sharply delimited algal patches, corticate; the cortex of rectangular to rounded cells, often with white hairs. *Ascocarp* an apothecium, immersed in the thallus, erumpent; proper exciple poorly developed; paraphyses almost unbranched, branched or anastomosing; epithecium containing algae; asci unitunicate, thick walled, 1 spored; ascospores muriform, hyaline, walls thin. *Pycnidia* not known.
CHEMISTRY: No data.
PHYCOBIONT: *Trebouxia*-like.
DISTRIBUTION: On leaves, in Queensland. 1 species reported (3).

GYMNODERMA NYL.: *Syn. Lich.* 26 (1860)
(including *Neophyllis* F. Wils.)
MORPHOLOGY: *Thallus* foliose to squamulose, upper surface corticate; lower surface of medullary hyphae, without rhizines. *Ascocarp* an apothecium on short podetia; podetia marginal on the lobes, hollow; apothecia capitate to barrel shaped; paraphyses reticulate, agglutinated; asci unitunicate, clavate, tholus I+, 8 spored; ascospores simple, hyaline, thin walled. *Pycnidia* immersed; fulcrum exobasidial.
CHEMISTRY: (3 species) atranorin (2), didymic acid (1), grayanic acid (1), protolichesterinic acid (1).
PHYCOBIONT: *Trebouxia.*
DISTRIBUTION: On bark, in humid forests of New South Wales and Victoria. 1 species reported (3).

GYMNOGRAPHA MULL. ARG: *Flora, Jena* 70, 62 (1887)

MORPHOLOGY: *Thallus* crustose, uniform, ecorticate; attached by prothallial or medullary hyphae. *Ascocarp* an apothecium, immersed, irregular, poorly developed or emarginate; hypothecium pale; paraphyses not known; asci 8 spored; ascospores transversely 3-septate, brown, thin walled. *Pycnidia* not known.
CHEMISTRY: No data.
PHYCOBIONT: *Trentepohlia.*
DISTRIBUTION: On bark, in Queensland. 1 species reported (1).

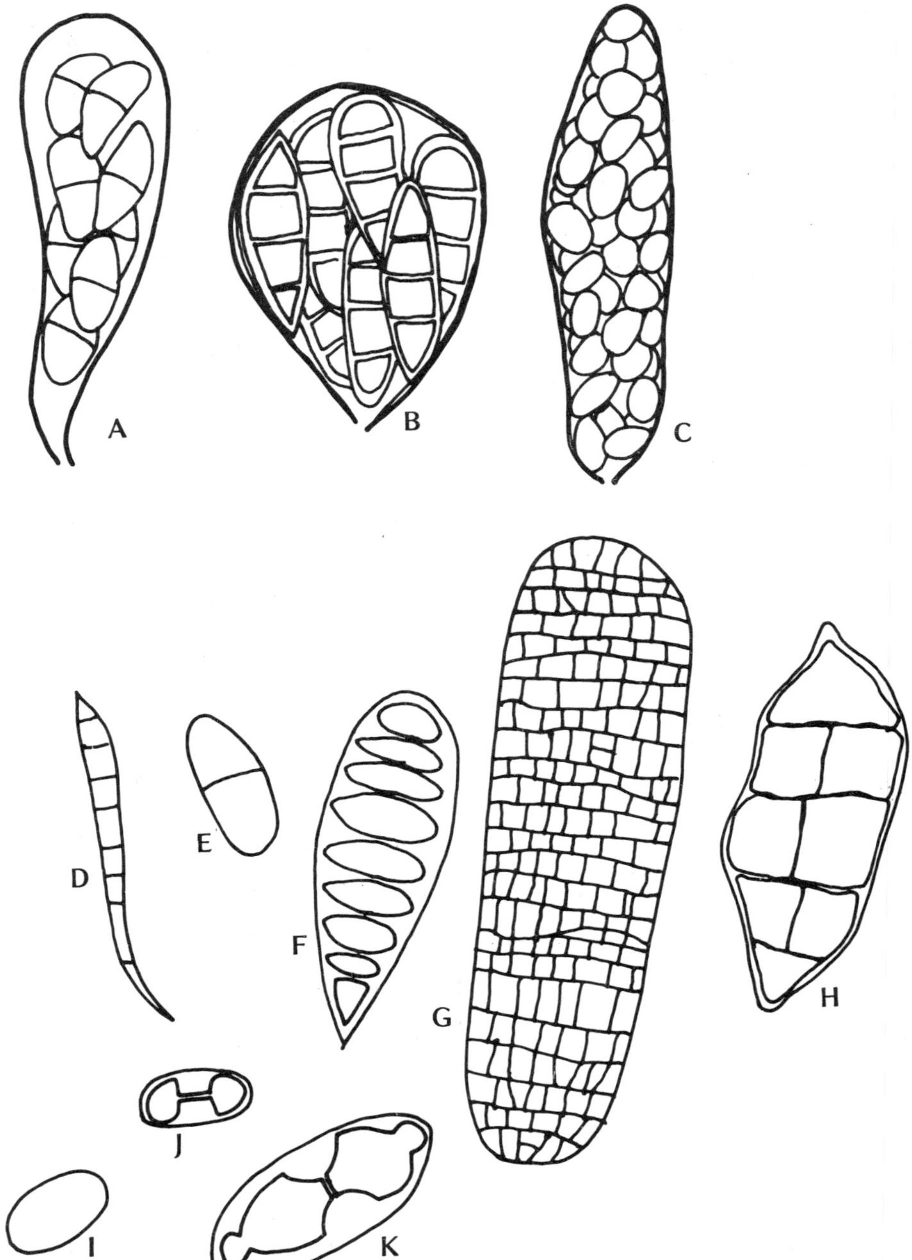

FIGURE 2 **Top:** A to C are asci containing spores: (A) *Buellia,* showing a clavate ascus with 1 septate spores, (C) *Acarospora* showing numerous minute simple spores. (B) shows a globuse ascus containing transversely septate spores of *Arthonia.*
Bottom: D to E are various spore types: (D) transversely septate thin walled spore of *Haematomma* showing cylindrical locules, (E) 1 septate thin walled spore of *Lopadium,* (F) transversely septate, unequally thickened spore of *Graphis,* showing lenticular locules, (G) muriform spore of *Graphina,* (H) muriform spore of *Diploschistes,* (I) simple spore of *Lecanora,* (J) polarilocular spore of *Caloplaca* showing the unequally thickened wall with two chambers joined by a canal, (K) 1 septate spore of *Buellia* showing unequal thickening and development of false septa near the ends.

HAEMATOMMA MASS.: *Ric. Lich. Crost.* 32 (1852)

MORPHOLOGY: Thallus crustose, uniform, corticate; cortex of agglutinate thin walled hyphae; attached to the substrate by medullary hyphae. *Ascocarp* an apothecium, immersed to sessile, round or somewhat irregular, thalloid margin usually well developed; hypothecium pale; paraphyses unbranched, branched, or reticulate; asci cylindrical, unitunicate, thin walled, tholus I+ blue, 8 spored; ascospores acicular, transversely 3-many septate, hyaline, thin walled. *Pycnidia* immersed; fulcrum endobasidial, little branched or unbranched; pycnospores bacilliform.
CHEMISTRY: (21 species) atranorin (10), thamnolic acid (6), divaricatic acid (3), psoromic acid (3), usnic acid (3), zeorin (3), porphyrilic acid (2), decarboxythamnolic acid (1), erythrommone (1), haemoventosin (1), imbricaric acid (1), pachycarpin (1), perlatolic acid (1), placodiolic acid (1), ventosic acid (1), xanthone pigments.
PHYCOBIONT: Coccoid Chlorophyceae.
DISTRIBUTION: On wood or bark, throughout the region. 6 species reported (30).

HELMINTHOCARPON FEE: *Essai Crypt. Suppl.* 156 (1837)

MORPHOLOGY: Thallus crustose, uniform, ecorticate or poorly corticate, attached to the substrate by medullary hyphae. *Ascocarp* an apothecium, apothecia immersed to adnate, irregular, roundish to stellate or branched; proper margin thin, usually covered in a thalloid layer; hypothecium pale; paraphyses thin, reticulate; asci cylindrical to clavate, unitunicate, I−, thin walled, 4-8 spored; ascospores fusiform to oblong, muriform, hyaline. *Pycnidia* not known.
CHEMISTRY: No data.
PHYCOBIONT: Trentepohlia.
DISTRIBUTION: Queensland. 2 species reported (15).

HEPPIA NAEG. in HEPP: *Flecht. Eur.* 49 (1853)

MORPHOLOGY: Thallus squamulose, paraplectenchymatous throughout, usually without a distinct medulla, attached to the substrate by hyphal rhizoids. *Ascocarp* an apothecium, immersed, disk open, round; proper exciple absent or dissappearing; paraphyses unbranched; hypothecium pale; asci clavate, unitunicate, thin walled, I+ blue, tholus I+ blue, 8 spored; ascospores simple, hyaline, thin walled. *Pycnidia* immersed, fulcrum exobasidial; pycnospores ellipsoid to oblong.
CHEMISTRY: No data.
PHYCOBIONT: Scytonema.
DISTRIBUTION: On soil, in arid areas. 1 species reported (1).

HERPOTHALLON TOBLER: *Flora, Jena* 131, 446 (1937)

MORPHOLOGY: Thallus of a loose construction, more or less crustose, forming patches up to 8 cm diameter, coloured red, white and green in concentric patterns; constructed from loosely woven hyphae which often terminate in pyramidal swellings. *Hymenium* not known.
CHEMISTRY: Not known.
PHYCOBIONT: Trentepohlia and some coccoid green algae.
DISTRIBUTION: In rainforests of Queensland, probably over mosses. 1 species reported (1).

HETERODEA NYL.: *Bull. Soc. Linn. Normandie ser 2,* 2, 47 (1867)

MORPHOLOGY: Thallus foliose, prostrate or ascending, much branched, upper surface corticate, cortex of agglutinated longitudinal hyphae; lower surface ecorticate, with a more or less complete layer of tomentum or rhizines which may appear cyphellate or veined. *Ascocarp* an apothecium, marginal, peltate, cymosely aggregate, proper exciple developed; hypothecium pale, over an algal layer; paraphyses unbranched; asci cylindrical, unitunicate, tholus I+, 8 spored; ascospores ovoid, simple, hyaline, thin walled. *Pycnidia* sessile, marginal; fulcrum endobasidial; pycnospores bacilliform.
CHEMISTRY: (2 species) usnic acid (2), condivaricatic acid (1), diffractaic acid (1), divaricatic acid (1).
PHYCOBIONT: Trebouxia-like.
DISTRIBUTION: Widespread in temperate and semi-arid Australia and New Caledonia. 2 species reported (2).

HETERODERMIA TREVIS.: *Atti Soc. Ital. Sci. Nat. Milano* 9, 613 (1868)

MORPHOLOGY: Thallus foliose, usually prostrate, attached by rhizines, lobes usually 2-5 mm broad, ciliate, with a well developed upper cortex of longitudinal hyphae; lower cortex often absent; medulla woolly. *Ascocarp* an apothecium, laminal; exciple thalloid; disk dark, often pruinose; paraphyses unbranched; asci unitunicate, clavate, thick walled, I+ blue, tholus I+ blue, 8 spored; ascospores 1 septate, brown, thick walled, with blastidia. *Pycnidia* immersed or prominent; fulcrum endobasidial; pycnospores bacilliform.
CHEMISTRY: (61 species) atranorin (61), zeorin (54), leucotylin (44), α-0-acetylleucotylins (43), triterpenoids (38), salazinic acid (25), norstictic acid (19), 2-chloroemodin (9), flavo-obscurins (8), 2-4 dichloroemodin (7), anthroquinones (5), dissectic acid (5).
PHYCOBIONT: Trebouxia.
DISTRIBUTION: On trees and rocks in Queensland, New South Wales, Victoria, Tasmania and South Australia. 12 species reported (60).

HYPOGYMNIA NYL.: *Lich. Env. Paris* 39 (1896)

MORPHOLOGY: Thallus foliose, radiate, corticate on both surfaces; cortex of anticlinal hyphae; medulla very lax, lobes often hollow, imperforate; without rhizines. *Ascocarp* an apothecium, stipitate, round, thalline exciple; disk brown; asci small, cylindrical, unitunicate, I+ blue, 8 spored; ascospores ellipsoid, simple, hyaline. *Pycnidia* immersed; fulcrum endobasidial; pycnospores bacilliform.
CHEMISTRY: (28 species) atranorin (28), physodic acid (24), chloroatranorin (20), physodalic acid (12), oxyphysodic acid (11), protocetraric acid (10), 2-0'-methylphysodic acid (6), salazinic acid (2), anthroquinones (1), barbatic acid (1), fumarprotocetraric acid (1), lobaric acid (1), usnic acid (1), vittatolic acid (1), zeorin (1).
PHYCOBIONT: Trebouxia.
DISTRIBUTION: Queensland, New South Wales, Victoria, Tasmania, South Australia and Western Australia in moist and cool to cold areas, commonly sub-alpine. 11 species reported (44).

HYPOTRACHYNA HALE: *Phytologia* 28, 340 (1974c)

MORPHOLOGY: Thallus foliose, lobes sub-linear, apically truncate, both surfaces corticate, paraplectenchymatous, attached by dichotomously branched rhizines, lower surface black.

Ascocarp an apothecium, adnate to substipitate, imperforate, round, thalloid exciple well developed; disk brown; hypothecium pale; paraphyses unbranched; asci cylindrical, unitunicate, thin walled, I+ blue, 8 spored; ascospores ovoid, simple, hyaline, thin walled. *Pycnidia* immersed; fulcrum endobasidial, bayonet-like; pycnospores bacilliform.
CHEMISTRY: (77 species) atranorin (57), protocetraric acid (15), usnic acid (14), obtusatic acid (13), barbatic acid (12), norstictic acid (11), lichexanthone (10), salazinic acid (10), rhodophyscin (9), alectoronic acid (8), colensoic acid (7), gyrophoric acid (7), lividic acid (6), anziaic acid (5), α-collatolic acid (5), entothein (4), evernic acid (4), lecanoric acid (4), constictic acid (3), echinocarpic acid (3), galbinic acid (3), sticic acid (3), olivetoric acid (2), physodalic acid (2), fumarprotocetraric acid (1), microphyllinic acid (1), perlatolic acid (1), succinprotocetraric acid (1).
PHYCOBIONT: Trebouxia.
DISTRIBUTION: On trees, especially in the southern half of the region. 8 species reported (83).

LAURERA REICHENBACH: *Deutsche Bot.* 15 (1841)

MORPHOLOGY: Thallus crustose, uniform, ecorticate; epi- or endophloic. *Ascocarp* a perithecium, immersed singly or aggregated in stroma; ostioles erect; paraphyses reticulate; asci cylindrical-clavate, I−; bitunicate, 2-8 spored; ascospores muriform, hyaline, thin walled. *Pycnidia* immersed; fulcrum exobasidial.
CHEMISTRY: (5 species) parietin (2), xanthorin (2), lichexanthone (1), papulosin (1).
PHYCOBIONT: Trentepohlia.
DISTRIBUTION: Queensland and Tasmania. 4 species reported (28).

LECANACTIS ESCHW.: *Syst. Lich.* 14 (1824)

MORPHOLOGY: Thallus crustose, uniform, ecorticate, attached by medullary or prothallial hyphae. *Ascocarp* a pseudothecium, immersed to sessile; circular or somewhat irregular; margin proper, black; hypothecium black; paraphyses branched, often reticulate; asci clavate, bitunicate, I−, 4-8 spored; ascospores oblong-acicular, transversely 3-15 septate, hyaline, thin walled. *Pycnidia* immersed; fulcrum exobasidial; pycnospores ellipsoid to oblong.
CHEMISTRY: (11 species) lecanoric acid (4), erythrin (3), phenolics (3), atranorin (1), conpsoromic acid (1), psoromic acid (1), schizopeltic acid (1), triterpenoids (1).
PHYCOBIONT: Trentepohlia.
DISTRIBUTION: Queensland and Tasmania. 5 species reported (90).

LECANIA MASS.: *Alcuni Gener. Lich.* 12 (1855)

MORPHOLOGY: Thallus crustose; uniform, lobed at the margins or squamulose; corticate or ecorticate; attached by medullary or prothallial hyphae; medulla woolly. *Ascocarp* an apothecium, sessile, round; margin thalloid; well developed, with or without a proper exciple; hypothecium pale; paraphyses unbranched; asci clavate, unitunicate, tholus I+ blue, 8 or rarely 16-32 spored; ascospores oblong-ellipsoid, 1 (rarely more) septate, hyaline, thin walled. *Pycnidia* immersed, fulcrum exobasidial.
CHEMISTRY: (1 species) lecanin.
PHYCOBIONT: Trebouxia.
DISTRIBUTION: On rock or bark, in Queensland, New South Wales, Victoria and Tasmania. 9 species reported (90).

LECANORA ACH.: *Lich. Univ.* 77 (1810)

Recent studies have resulted in a number of genera being segregated from this taxon (Hawksworth, James and Coppins 1980). Knowledge of Australian material is insufficient to allow certainty about which are present in Australia.

MORPHOLOGY: *Thallus* crustose, uniform or lobed at the margins; ecorticate or with a developed cortex; medulla woolly; attached by medullary or prothallial hyphae. *Ascocarp* an apothecium, round, sessile; thalloid margin well developed; paraphyses unbranched; hypothecium pale, asci clavate, unitunicate, tholus I+ blue, 8 (or rarely 16-32) spored; ascospores oblong-ellipsoid simple, hyaline, thin walled. *Pycnidia* immersed; fulcrum usually exobasidial, rarely endobasidial; pycnospores bacilliform to filiform.
CHEMISTRY: (53 species) atranorin (20), usnic acid (19), zeorin (7), chloroatranorin (6), norstictic acid (5), lecanoric acid (4), protocetraric acid (3), psoromic acid (3), rocellic acid (3), epanorin (2), gangaleodin (2), rhizocarpic acid (2), α-collatolic acid (1), divaricatic acid (1), fumarprotocetraric acid (1), leucotylin (1), lobaric acid (1), melanophthalmic acid (1), olivetoric acid (1), pannarin (1), placodiolic acid (1), sordidone (1), sulphurellin (1), thiophanic acid (1), variolaric acid (1).
PHYCOBIONT: *Trebouxia.*
DISTRIBUTION: On trees, soil and rock, throughout Australia. 70 species reported (400).

LECIDEA ACH. em ZAHLBR.: *Nat. Pflanz.* 1, 130 (1905)

Recent studies have resulted in a number of genera being segregated from this taxon (Hawksworth, James and Coppins 1980). Knowledge of Australian material is insufficient to allow certainty about which are present in Australia.

MORPHOLOGY: *Thallus* crustose; uniform, warty-cracked areolate to marginally lobed; ecorticate or corticate; attached to the substrate by medullary or prothallial hyphae. *Ascocarp* an apothecium, sessile, disk round; margin proper; hypothecium hyaline, coloured or black; paraphyses unbranched, often capitate at the tip; asci cylindrical-clavate, unitunicate, tholus I+ blue, 8 spored; ascospores ellipsoid, simple, hyaline, thin walled. *Pycnidia* immersed; fulcrum exobasidial; pycnospores bacilliform to filiform.
CHEMISTRY: (65 species) gyrophoric acid (12), atranorin (11), lecanoric acid (8), norstictic acid (8), confluentic acid (4), fumarprotocetraric acid (4), constictic acid (2), lichexanthone (2), 2-0-methylanziaic acid (2), planaic acid (2), psoromic acid (2), rhizocarpic acid (2), salazinic acid (2), stictic acid (2), anziaic acid (1), arthothelin (1), calycin (1), chloroatranorin (1), 7-chloroemodin (1), diploicin (1), divaricatic acid (1), ergosterol (1), glomelliferic acid (1), leprapinic acid (1), 2-0-methylconfluentic acid (1), 2-0-methylperlaolic acid (1), miriquidic acid (1), norsolorinic acid (1), perlatolic acid (1), porphyrilic acid (1), thiophanic acid (1),thuringione (1), schizopeltic acid (1).
PHYCOBIONT: *Trebouxia, Chlorella, Chlorosarcina, Coccobotrys, Myrmecia, Pleurococcus, Pseudochlorella.*
DISTRIBUTION: On trees, rocks and soil, throughout Australia. 87 species reported (800).

LEMPHOLEMMA KORB.: *Syst. Lich. Germ.* 400 (1855)

MORPHOLOGY: *Thallus* wholly mycelial, irregular, gelatinous, lobed or branched, smooth or rough, taking the form of the phycobiont; ecorticate; attached to the substrate by hyphae. *Ascocarp* an apothecium, immersed to adnate; sometimes lateral or terminal; disk round, flat

to concave; margin thalloid, thin; hypothecium hyaline; paraphyses unbranched; asci clavate, unitunicate, thin walled, I+ blue, 8 spored; ascospores spherical to ellipsoid, simple, hyaline, thin walled. *Pycnidia* immersed; fulcrum exobasidial; pycnospores bacilliform.
CHEMISTRY: No data.
PHYCOBIONT: *Nostoc.*
DISTRIBUTION: Queensland and Victoria, in humid forests. 2 species reported (30).

LEPRARIA ACH.: *Meth. Lich.* 3 (1803)

MORPHOLOGY: *Thallus* crustose, devoid of differentiation, uniform, of tangled hyphae and phycobiont cells, forming a sorediod mass. *Ascocarp* not known. *Pycnidia* not known.
CHEMISTRY: (10 species) atranorin (4), calycin (2), fumarprotocetraric acid (2), leprapinic acid (2), roccellic acid (2), barbatolic acid (1), divaricatic acid (1), lepraric acid (1), pinastric acid (1), usnic acid (1), vulpinic acid (1), zeorin (1).
PHYCOBIONT: *Chlorella, Stichococcus, (+?Trebouxia).*
DISTRIBUTION: On rock, wood and bark, throughout the region. 2 species reported (20).

LEPROCAULON NYL. ex LAMY: *Bull. Soc. Bot. Fr.* 25, 352 (1868)

MORPHOLOGY: *Thallus* fruticose, dwarfed, ecorticate, smooth, branches slender, more or less erect, sometimes squamulose, becoming powdery, fragile. *Ascocarp* not known. *Pycnidia* not known.
CHEMISTRY: (7 species, all variable) atranorin (5), rangiformic acid (4), fatty acids (4), baeomycesic acid (3), conpsoromic acid (3), protocetraric acid (3), psoromic acid (3), divaricatic acid (2), barbatic acid (1), dibenzofuran (1), didymic acid (1), evernic acid (1), grayanic acid (1), lecanoric acid (1), physodalic acid (1), sekikaic acid (1), succinprotocetraric acid (1), thamnolic acid (1), usnic acid (1), zeorin (1).
PHYCOBIONT: *Trebouxia.*
DISTRIBUTION: On dry rock ledges, in New South Wales, Victoria, Tasmania and South Australia. 2 species reported (7).

LEPTOGIUM (ACH.) GRAY: *Nat. Arr. Br. Pl.* 400 (1821)

MORPHOLOGY: *Thallus* foliose to squamulose or fruticose; membranous, gelatinous; corticate, with two cell layers; medulla of hyphae in algal gel; adhering to the substrate or umbilicate. *Ascocarp* an apothecium, initially immersed, later adnate to stipitate; laminal or marginal; disk circular, margin thalline; hypothecium hyaline to brown, of interwoven hyphae or pseudoparenchymatous; paraphyses unbranched; asci clavate, unitunicate, tholus I+ blue, 8 spored; ascospores oval to ellipsoid, hyaline, muriform, thin walled. *Pycnidia* immersed in the thallus or in warts; fulcrum endobasidial, small celled; pycnospores bacilliform.
CHEMISTRY: (21 species) no secondary lichen substances found.
PHYCOBIONT: *Nostoc.*
DISTRIBUTION: On rock, bark and soil, in humid areas of Queensland, New South Wales, Victoria, Tasmania, South Australia and Western Australia. 30 species reported (50).

LEPTOTREMA MONT. et BOSCH: *Plant. Junghuhn.* 4, 483 (1855)

MORPHOLOGY: *Thallus* crustose, epi- or endophloic, uniform; ecorticate or with a woolly poorly developed cortex, attached by medullary or prothallial hyphae. *Ascocarp* an apothecium,

immersed to sessile in warts; disk round or elongate, punctiform or narrow; well developed proper exciple initially closed over the disk, later tearing and opening, then disintegrating leaving the disk free but surrounded by a thalline margin; paraphyses unbranched; asci cylindrical-clavate, unitunicate, thick walled, I−, 8 spored; ascospores ellipsoid or oblong-ellipsoid, brown, muriform, wall thickness unequal resulting in spherical or lenticular locules. *Pycnidia* fulcrum exobasidial (?).
CHEMISTRY: (17 species) stictic acid (7), constictic acid (5), norstictic acid (2), protocetraric acid (2), conpsoromic acid (1), hypoprotocetraric acid (1), psoromic acid (1), virensic acid (1).
PHYCOBIONT: *Trentepohlia.*
DISTRIBUTION: Queensland and Tasmania, in humid forests. 12 species reported (65).

LICHINA C. AG.: *Spec. Alg.* 104 (1821)

MORPHOLOGY: *Thallus* minutely fruticose, caespitose, gelatinous; branches cylindrical or flattened with a cortex of tangled hyphae; medulla of longitudinal hyphae. *Ascocarp* an apothecium, immersed in the swollen tips of branches; disk round or irregular; narrow proper exciple enclosed in a thalline exciple; paraphyses unbranched (rarely branched), thin; asci cylindrical, unitunicate, thin walled, 8 spored; ascospores ellipsoid, simple, hyaline, thin walled. *Pycnidia* immersed in the thallus close to the apothecia; fulcrum exobasidial; pycnospores bacilliform.
CHEMISTRY: (2 species) no secondary lichen substances.
PHYCOBIONT: *Calothrix.*
DISTRIBUTION: On maritime rocks, in the southern parts of Australia. 2 species reported (7).

LOBARIA SCHREB.: *Gener. Plant.* II, 768 (1791)

MORPHOLOGY: *Thallus* foliose, prostrate or ascending, attached to the substrate by a bushy tomentum; corticate on both surfaces, both cortex layers paraplectenchymatous, medulla woolly. *Ascocarp* an apothecium, marginal or laminal, sessile to stipitate; initially urn-like, later more open, disk concave, exciple paraplectenchymatously corticate, thalline; hypothecium pale or brown; paraphyses unbranched; asci cylindrical-clavate, unitunicate, tholus I+ blue, 8 spored; ascospores fusiform-acicular, transversely 1-several septate, hyaline, thin walled. *Pycnidia* in small warts; fulcrum endobasidial, small celled; pycnospores bacilliform or somewhat constricted at the centre.
CHEMISTRY: (55 species) gyrophoric acid (35), norstictic acid (15), stictic acid (14), 4-0-methylgyrophoric acid (13), constictic acid (10), thelephoric acid (9), atranorin (8), triterpenes (6), tenuiorin (4), scrobiculin (3), usnic acid (2), retigeric acid (1).
PHYCOBIONT: *Myrmecia, Trebouxia, Nostoc* (*Nostoc* often in cephalodia).
DISTRIBUTION: On bark and rock, in Queensland, New South Wales, Victoria and Tasmania, mostly in wetter areas. 22 species reported (80).

LOPADIUM KORB.: *Syst. Lich. Germ.* 210 (1855)

MORPHOLOGY: *Thallus* crustose, uniform ecorticate, attached to the substrate by prothallial or medullary hyphae. *Ascocarp* an apothecium, sessile or stipitate; disk round, margin soft or cartilaginous, proper; hypothecium pale to dark; hymenium mucilagenous; paraphyses unbranched or branched; asci cylindrical-clavate, unitunicate, thick walled, tholus I+ blue, 1-8 spored; ascospores ovoid, muriform, hyaline, thin walled. *Pycnidia* sessile; fulcrum endobasidial; pycnospores bacilliform.

CHEMISTRY: (8 species) 2-7 dichlorolichexanthone (4), parietin (3), atranorin (2), chloroatranorin (1), triterpene (1), usnic acid (1).
PHYCOBIONT: *Trebouxia.*
DISTRIBUTION: Queensland, New South Wales and Tasmania, on bark in humid forests. 13 species reported (70).

MALLOTIUM GRAY: *Nat. Arr. Br. Pl.* 1, 399 (1821)

MORPHOLOGY: *Thallus* foliose, large; upper surface with a single layer of paraplectenchymatous cells; lower surface covered with a tomentum of filaments; gelatinous when wet. *Ascocarp* an apothecium, initially immersed, later adnate to stipitate; disk circular, margin thalline; hypothecium hyaline to brown; paraphyses unbranched; asci clavate, unitunicate, tholus I+ blue, 8 spored; ascospores oval to ellipsoid, muriform, hyaline, thin walled. *Pycnidia* immersed in the thallus or warts; fulcrum endobasidial; pycnospores bacilliform.
CHEMISTRY: No secondary lichen substances known.
PHYCOBIONT: *Nostoc.*
DISTRIBUTION: Victoria and Western Australia. 3 species reported (12).

MARONEA MASS.: *Misc. Lich.* 291 (1856)

MORPHOLOGY: *Thallus* crustose, uniform; ecorticate or partly corticate; medulla woolly; attached by medullary hyphae. *Ascocarp* an apothecium, at first immersed, later sessile; disk round; margin thalloid only, corticate; hypothecium pale; paraphyses unbranched or little branched; asci clavate-globular, unitunicate, thin walled, I+ pale blue, tholus I+ pale blue, many spored; ascospores globular, simple (but sometimes apparently septate), thin walled, hyaline. *Pycnidia* immersed; fulcrum exobasidial; pycnospores thread-like, cylindrical.
CHEMISTRY: No data.
PHYCOBIONT: *Trebouxia.*
DISTRIBUTION: Victoria and South Australia, on bark. 1 species reported (13).

MASTODIA HOOK. f. et HARV. apud HOOK. f.: *Bot. Antarctic Voyage* 1,499 (1847)

MORPHOLOGY: *Thallus* foliose to squamulose, of loosely woven hyphae and algae; without rhizines or prothallus. *Ascocarp* a perithecium, immersed in the thallus; paraphyses gelatinizing; ascus 8 spored; ascospores elongate, simple, hyaline. *Pycnidia* immersed; fulcrum exobasidial, pycnospores cylindrical, oblong to ellipsoid.
CHEMISTRY: No data.
PHYCOBIONT: *?Prasiola.*
DISTRIBUTION: Tasmania. 1 species reported (7).

MAZOSIA MASS.: *Neag. Lich.* 9 (1854a)

MORPHOLOGY: *Thallus* crustose, uniform, ecorticate; attached to the substrate by medullary or prothallial hyphae. *Ascocarp* a pseudothecium; round or irregular; immersed in the thallus; margin rudimentary to well developed; hypothecium hyaline to black; paraphyses reticulate; asci clavate, bitunicate, thin, I−, 8 spored; ascospores acicular to ovate, hyaline, transversely many septate, thin walled. *Pycnidia* immersed; fulcrum exobasidial; pycnospores cylindrical, oblong to ellipsoid.

CHEMISTRY: (3 species) No secondary lichen substances.
PHYCOBIONT: *Phycopeltis.*
DISTRIBUTION: Humid forests in Queensland, normally foliicolous. 3 species Reported (9).

MEDUSULINA MULL. ARG.: *Bull. Herb. Boissier* 2, 93 (1894a)

MORPHOLOGY: *Thallus* crustose, uniform, ecorticate, attached to the substrate by hyphal rhizoids. *Ascocarp* an apothecium, lirreliform, grouped in raised stroma, immersed to superficial; exciple well developed; hypothecium hyaline; paraphyses unbranched; ascus cylindrical to clavate, unitunicate, I−, 1-8 spored, ascospores oblong-ellipsoid to fusiform, hyaline, muriform, walls unequally thickened resulting in spherical or lenticular locules. *Pycnidia* not known.
CHEMISTRY: (2 species) norstictic acid (1), stictic acid (1).
PHYCOBIONT: *Trentepohlia.*
DISTRIBUTION: Humid forests in Queensland and Victoria, on bark. 2 species reported (6).

MEGALOSPORA MEYEN et FLOTOW: *Nov. acta Acad. Caesar Leop. Carol. 19 Suppl.* 228 (1843)

MORPHOLOGY: *Thallus* crustose, uniform, ecorticate, often sorediate; attached to the substrate by medullary or prothallial hyphae. *Ascocarp* an apothecium, sessile or adnate; disk circular, flat or convex; margin proper; hypothecium pale; paraphyses unbranched, branched or anastomosing; asci clavate, unitunicate, I+ blue, tholus I+ blue, 1-8 spored; ascospores ellipsoid, 1 septate, hyaline, thick walled. *Pycnidia* immersed; fulcrum endobasidial; pycnospores bacilliform.
CHEMISTRY: (4 species) usnic acid (4), zeorin (4).
PHYCOBIONT: *Trebouxia.*
DISTRIBUTION: Humid forests, in Queensland, Victoria, New South Wales and Tasmania. 12 species reported (50).

MELAMPYDIUM STIRT.: *J. Linn. Soc. Bot.* 16, 471 (1875)

MORPHOLOGY: *Thallus* crustose, uniform, ecorticate; attached to the substrate by medullary or prothallial hyphae. *Ascocarp* an apothecium-like pseudothecium; round to irregular; immersed or sessile, flat; proper margin little developed; hypothecium pale; paraphyses little branched; asci clavate, bitunicate, thin walled, I−, 8 spored; ascospores transversely septate or muriform, hyaline, thin walled. *Pycnidia* not known.
CHEMISTRY: No data.
PHYCOBIONT: *Trebouxia*-like.
DISTRIBUTION: Queensland and Tasmania, on bark. 1 species reported (1).

MELANELIA ESSL.: *Mycotaxon* 7, 46 (1978)

MORPHOLOGY: *Thallus* foliose, lobed, radiate; loosely to moderately adnate to the substrate, attached by rhizines; lobes short and rotund or elongate, more or less flat; with a paraplectenchymatous cortex; commonly sorediate-isidiate or pseudocyphellate above, HNO_3−. *Ascocarp* an apothecium, laminal, sessile; disk concave, round; margin thalloid; hypothecium pale; paraphyses branched; ascus clavate, unitunicate, I+ blue, tholus I+ blue,

8 spored; ascospores ellipsoid simple, hyaline, thin walled. *Pycnidia* immersed; fulcrum endobasidial; pycnospores bacilliform.
CHEMISTRY: (37 species) lecanoric acid (8), gyrophoric acid (7), protocetraric acid (6), fumarprotocetraric acid (5), perlatolic acid (4), stenosporic acid (4), rhodophyscin (2), caperatic acid (1), salazinic acid (1).
PHYCOBIONT: *Trebouxia.*
DISTRIBUTION: Throughout the southern part of the region, on a range of substrates. 4 species reported (37).

MELANOTHECA FEE: *Essai Crypt. Suppl.* 70 (1837)

MORPHOLOGY: *Thallus* crustose, uniform, ecorticate, epi- or endophloic. *Ascocarp* a perithecium, immersed in groups in stroma, black; ostiole straight, erect; paraphyses unbranched or reticulate; asci cylindrical-clavate, bitunicate, thin walled, I−, 8 spored; ascospores ellipsoid to fusiform, transversely 3-several septate, brown, with unequally thickened walls resulting in lenticular or globose locules. *Pycnidia* immersed; fulcrum exobasidial; pycnospores filiform, curved.
CHEMISTRY: No data.
PHYCOBIONT: *Trentepohlia.*
DISTRIBUTION: Queensland and New South Wales. 6 species reported (53).

MELASPILEA NYL.: *Act. Soc. Linn. Bordeaux* 21, 416 (1856)

MORPHOLOGY: *Thallus* crustose, uniform, ecorticate; epi- or endophloic; attached to the substrate by prothallial or medullary hyphae. *Ascocarp* an apothecium, round or irregular to elongate; devoid of thalloid margin; proper exciple often black; hypothecium hyaline to brown; paraphyses unbranched or absent; asci cylindrical, unitunicate, thin walled, I−, 8 spored, ascospores ellipsoid or asymmetric, 1 septate, hyaline or brown, thin walled. *Pycnidia* immersed; fulcrum exobasidial; pycnospores bacilliform.
CHEMISTRY: (1 species) no secondary lichen substances found.
PHYCOBIONT: *Trentepohlia.*
DISTRIBUTION: Queensland, New South Wales, Victoria and Tasmania. 9 species reported (60).

MENEGAZZIA MASS.: *Neag. Lich.* 3 (1854a)

MORPHOLOGY: *Thallus* foliose, radiately lobed to sub-fruticose; corticate on both surfaces; upper surface perforate; lower surface black, without rhizines; medulla very lax, normally hollow. *Ascocarp* an apothecium, sessile to stipitate; round, well developed thalloid margin; hypothecium hyaline; paraphyses unbranched; asci cylindrical-clavate, unitunicate, I+, tholus I+ blue, usually 2-4 spored; ascospores ellipsoid, simple, hyaline, thick walled, often large. *Pycnidia* immersed; fulcra endobasidial, bayonet-like; pycnospores bacilliform.
CHEMISTRY: (3 species) atranorin (3), stictic acid (3), constictic acid (1).
PHYCOBIONT: *Trebouxia.*
DISTRIBUTION: Widely distributed in the southern half of the continent especially in cooler and wetter areas. 12 species reported (30).

MICAREA FR. em HEDL.: *Bih. Kgl. Svensk Vetensk Akad. Handl.* 18(3) 3, 27 (1892)

MORPHOLOGY: *Thallus* crustose, uniform; composed at least in part of goniocyste-like granules. *Ascocarp* an apothecium, sessile, emarginate, often globose; hypothecium pale or dark; paraphyses reticulate; asci unitunicate (?), thick walled, I+ blue, tholus I+ blue, 8 spored; ascospores ellipsoid, simple or transversely septate, hyaline. *Pycnidia* not known.
CHEMISTRY: No data.
PHYCOBIONT: *?Myrmecia.*
DISTRIBUTION: Tasmania and probably other regions, on acid substrates. 1 species reported (30).

MICROTHELIA KORB.: *Syst. lich. Germ.* 372 (1855)

MORPHOLOGY: *Thallus* crustose, uniform, granular, ecorticate; epi- or endophloic, with an indistinct prothallus. *Ascocarp* a perithecium, sessile or partly immersed, black, ostiole erect; paraphyses reticulate, often gelatinizing; asci cylindrical-clavate, bitunicate, thick walled, I−, 4-8 spored; ascospores ellipsoid to fusiform, usually 1 septate but up to 5 septate, brown, thin walled. *Pycnidia* immersed; fulcrum exobasidial; pycnospores bacilliform.
CHEMISTRY: No data.
PHYCOBIONT: *Trentepohlia.*
DISTRIBUTION: Queensland, New South Wales and Victoria. 10 species reported (70).

MYCOPORELLUM MULL. ARG.: *Rev. Mycol.* 6, 14 (1884)

MORPHOLOGY: *Thallus* crustose, uniform, ecorticate; epi- or endophloic. *Ascocarp* a perithecium-like pseudothecium, opening by a pore or irregular slit, black; often several immersed in a stroma; paraphyses absent or reticulate; asci oval or clavate, bitunicate, thick walled, I−, 8 spored; ascospores ellipsoid or asymetric to acicular, transversely 1-5 septate, hyaline or dark, thin walled, the locules often markedly unequal. *Pycnidia* not known.
CHEMISTRY: No data.
PHYCOBIONT: *Trentepohlia.*
DISTRIBUTION: Queensland, Victoria and Tasmania. 3 species reported (30).

NEOFUSCELIA ESSL.: *Mycotaxon* 7, 49 (1978)

MORPHOLOGY: *Thallus* foliose to almost crustose or sub-fruticose; loosely or closely adnate, attached by rhizines; lobes short and rotund to linear-elongate; more or less flat to convex; with a paraplectenchymatous cortex; with neither soredia nor isidia; HNO_3+. *Ascocarp* an apothecium, laminal, sessile; disk concave, round; margin thalloid; hypothecium pale; paraphyses branched; ascus clavate, unitunicate, I+ blue, tholus I+ blue, 8 spored; ascospores ellipsoid, simple, hyaline, thin walled. *Pycnidia* immersed; fulcrum endobasidial; pycnospores bacilliform.
CHEMISTRY: (67 species) norstictic acid (12), gyrophoric acid (8), divaricatic acid (6), glomelliferic acid (6), perlatolic acid (6), fumarprotocetraric acid (5), glomellic acid (5), physodic acid (5), protocetraric acid (5), stenosporic acid (5), alectoronic acid (4), α-collatolic acid (4), salazinic acid (4), diffractaic acid (3), barbatic acid (2), constictic acid (2), loxodellic acid (2), norlobaridone (2), olivetoric acid (2), stictic acid (2), caperatic acid (1), 4-0-methylbarbatic acid (1), 4-0-methylolivetoric acid (1), 4-0-methylphysodic acid (1).
PHYCOBIONT: *Trebouxia.*

DISTRIBUTION: Throughout the southern part of the region, on rocks and soil. 10 species reported (67).

NEPHROMA ACH.: *Lich. Univ.* 101 (1810)

MORPHOLOGY: Thallus foliose, large, radiate, prostrate or a little ascending; both surfaces paraplectenchymatously corticate; attached to the substrate by rhizoids. *Ascocarp* an apothecium, immersed on the lower surface of lobes which recurve; without an exciple; hypothecium pale; paraphyses unbranched; asci clavate, unitunicate, I−, 8 spored; ascospores fusiform, transversely 1-5 septate, brown, thin walled. *Pycnidia* marginal, immersed in small warts; fulcrum endobasidial, branched; pycnospores short, straight, constricted in the middle.
CHEMISTRY: (11 species) nephrin (6), zeorin (6), usnic acid (2), nephroarctin (2), nephromin (2), anthraquinones (1), emodin (1), fragilin (1), phenarctin (1), stictic acid (1).
PHYCOBIONT: Coccomyxa or *Nostoc.*
DISTRIBUTION: Queensland, New South Wales, Victoria and Tasmania. 7 species reported (35).

NEUROPOGON NEES et FLOTOW: *Linnaea* 9, 496 (1835)

MORPHOLOGY: Thallus fruticose, erect or somewhat pendulous, corticate; cortex horny, brittle, of agglutinate periclinal hyphae; medulla woolly, central axis cartilagenous, solid, of longitudinal agglutinated hyphae; attached to the substrate by a basal disk; branches yellow, black at the tips, glossy. *Ascocarp* an apothecium, lateral or terminal on branches, well developed thalline margin; disk round, black or dark; hypothecium dark; paraphyses branched, agglutinate; asci clavate to cylindrical, unitunicate, thick walled, I+ blue, tholus I+ blue, 8 spored; ascospores ellipsoid to spherical, simple, hyaline, thin walled. *Pycnidia* immersed; fulcrum exobasidial; pycnospores fusiform to acicular.
CHEMISTRY: (5 species) usnic acid (5), norstictic acid (2), fumarprotocetraric acid (1), salazinic acid (1).
PHYCOBIONT: Trebouxia.
DISTRIBUTION: Alpine or sub-alpine areas of Victoria and Tasmania. 5 species reported (10).

NORMANDINA VAIN.: *Acta Soc. Fauna Flora Fenn.* 7(2), 188 (1890)

MORPHOLOGY: Thallus squamulose to foliose; round to somewhat lobed; apressed or a little ascending; poorly developed upper cortex; attached by hyphal rhizoids. *Ascocarp* not known. *Pycnidia* not known.
CHEMISTRY: No identified secondary lichen substances.
PHYCOBIONT: Coccoid Chlorophyceae.
DISTRIBUTION: Queensland, New South Wales, Victoria and South Australia. 1 species reported (1).

OCELLULARIA MEYEN et FLOTOW: *Nov. Act. Acad. Caesar. Leop. Carol. 19 Supp.* (1843)

MORPHOLOGY: Thallus crustose, uniform; ecorticate or with a poorly developed cortex; attached to the substrate by medullary or prothallial hyphae. *Ascocarp* an apothecium,

immersed in the thallus or sessile in warts; disk round or rarely elongate; concave or flat; proper exciple completely closing the disk, but cracking with age and the remnants bordering the disk, or the disk left free within a thalloid margin; hypothecium hyaline; paraphyses unbranched; asci clavate, unitunicate, thick walled, I− (but hymenial gel may be I+), 1-8 spored; ascospores oblong to fusiform, transversely 1-many septate, hyaline, walls unequally thickened to form lenticular locules. *Pycnidia* unknown.
CHEMISTRY: (53 species) hypoprotocetraric acid (19), psoromic acid (12), conpsoromic acid (10), 4-0-demethylnotatic acid (10), protocetraric acid (5), constictic acid (4), stictic acid (4), endocrin (2), gyrophoric acid (1), lichexanthone (1), 4-0-methylhypoprotocetraric acid (1), norstictic acid (1), triterpenoids (1).
PHYCOBIONT: *Trentepohlia.*
DISTRIBUTION: Queensland, New South Wales and Victoria. 22 species reported (120).

OCHROLECHIA MASS.: *Ric. Lich. Crost.* 30 1852)

MORPHOLOGY: *Thallus* crustose, uniform, areolate, the areoles sometimes dwarf-fruticose; usually ecorticate, sometimes with a cortex of anticlinal hyphae; medulla woolly; attached to the substrate by medullary or prothallial hyphae. *Ascocarp* an apothecium, at first immersed later sessile and constricted at the base; disk round; margin thalloid; hypothecium pale; paraphyses reticulate; asci clavate, unitunicate (but of several layers), thick walled, I+ blue, tholus I+, 2-8 spored; ascospores ellipsoid, simple, hyaline, walls thin or little thickened. *Pycnidia* immersed in warts; fulcrum exobasidial; pycnospores oblong to cylindrical.
CHEMISTRY: (29 species) lecanoric acid (18), gyrophoric acid (10), erythrin (5), variolaric acid (4), atranorin (3).
PHYCOBIONT: *Trebouxia.*
DISTRIBUTION: Queensland, New South Wales, Victoria, Tasmania and South Australia. 6 species reported (40).

OPEGRAPHA HUMB.: *Flor. Friberg.* 57 (1793)

MORPHOLOGY: *Thallus* crustose, uniform, ecorticate, sometimes endophloic; attached by medullary or prothallial hyphae. *Ascocarp* a lirreliform pseudothecium, immersed, adnate or sessile, round or markedly elongate, with carbonaceous proper exciple; hypothecium dark or pale; paraphyses reticulate; asci clavate, bitunicate, thin walled, I−, 8 spored; ascospores ellipsoid to fusiform, transversely 1-many septate, hyaline or brown, thin walled. *Pycnidia* immersed; fulcrum exobasidial; pycnospores oblong to filiform, curved.
CHEMISTRY: (8 species) erythrin (3), lecanoric acid (3), confluentic acid (1), farinosin (1), norstictic acid (1), triterpenes (1).
PHYCOBIONT: *Trentepohlia, Phycopeltis.*
DISTRIBUTION: Queensland, Victoria and Tasmania. 24 species reported (300).

PANNARIA DEL. in BORY.: *Dict. Class. hist. nat.* 20 (1828)

MORPHOLOGY: *Thallus* squamulose to foliose; upper surface corticate, of anticlinal hyphae forming paraplectenchyma; attached to the substrate by a well developed blue to black hypothallus or by a tomentum of dark rhizoids; lower surface ecorticate, of loosely woven hyphae; medulla at least in part woolly. *Ascocarp* an apothecium, at first immersed, eventually sessile; laminal; disk circular; thalloid exciple; hypothecium hyaline or pale coloured; paraphyses unbranched or branched; ascus clavate, unitunicate, I+, tholus I+, 8 spored;

ascospores ellipsoid to fusiform, simple, hyaline, sometimes the wall thickened or warty. *Pycnidia* immersed in thalline warts; fulcrum endobasidial; pycnospores bacilliform.
CHEMISTRY: (6 species) pannarin (3), atranorin (2), triterpenoids (2).
PHYCOBIONT: *Nostoc, Scytonema.*
DISTRIBUTION: Widespread in moist areas of all states, on rock, soil and bark. 18 species reported (80).

PANNOPARMELIA (MULL. ARG.) DARB.: *Wiss. Ergebn. Schwedisch Sudpolar-Expedit.* 1901-3 vol. 4(2), 11 (1912)

MORPHOLOGY: *Thallus* foliose, much divided, often moniliform; upper surface yellow-green, corticate, of anticlinal hyphae; lower cortex with a dense tomentum of black anastomosing hyphae in discrete patches. *Ascocarp* an apothecium, laminal, circular, stipitate; margin thalloid; hypothecium pale; paraphyses unbranched or little branched; asci cylindrical, unitunicate, I+ blue, tholus I+ blue, 8 spored; ascospores ellipsoidal to almost globose, simple, hyaline, thin walled.
CHEMISTRY: No data.
PHYCOBIONT: *Trebouxia.*
DISTRIBUTION: Queensland, New South Wales, Victoria and Tasmania. 3 species reported (5).

PARATHELIUM NYL.: *Bot. Ztg.* 20, 279 (1862)

MORPHOLOGY: *Thallus* crustose, uniform, epi-or endophloic, ecorticate. *Ascocarp* a perithecium; ostiolar canal more or less lateral; sometimes with a thalloid covering; paraphyses reticulate; asci cylindrical-clavate, bitunicate, usually thick walled, I−, 4-8 spored; ascospores ellipsoid to circular, transversely 3-10 septate, thin walled, brown. *Pycnidia* not known.
CHEMISTRY: No data.
PHYCOBIONT: *Trentepohlia.*
DISTRIBUTION: Queensland, on bark in humid forests. 1 species reported (?15).

PARMELIA ACH. *(sensu ampl.):* *Meth. Lich.* 153 (1803)
This very large generic aggregate is the subject of controversy. It has been divided by Hale (1974,a,b,c,d) and Esslinger (1978) into a great many segregates, which other workers may not accept. The segregates have been treated separately here, and may be found under the following names: *Bulbothrix, Hypotrachyna, Melanelia, Neofuscelia, Parmelia, Parmelina, Parmotrema, Pseudoparmelia, Relicina* and *Xanthoparmelia.*

PARMELIA ACH.: *(sensu Hale).*

MORPHOLOGY: *Thallus* foliose, lobed, radiate, appressed to somewhat ascending, both surfaces corticate; upper cortex of anticlinal hyphae, reticulately cracked or maculate; medulla woolly; lower cortex pale or dark, with simple or squarrose rhizines to the margin; with or without marginal cilia. *Ascocarp* an apothecium, sessile, laminal, disk concave, brown, circular; margin thalloid; hypothecium pale; paraphyses branched; ascus clavate, unitunicate, I+ blue, tholus I+ blue, 8 spored; ascospores ellipsoid, simple, hyaline, thin walled. *Pycnidia* immersed; fulcrum endobasidial, bayonet-like; pycnospores bacilliform.
CHEMISTRY: (55 species) salazinic acid (28), lecanoric acid (10), gyrophoric acid (5), all species probably contain atranorin.

PHYCOBIONT: *Trebouxia.*
DISTRIBUTION: Widespread throughout the region except in arid or semi-arid areas. ?8 species reported (?60).

PARMELIELLA MULL. ARG.: *Mem. Soc. Phys. Hist. nat. Geneve* 16(2), 376 (1862).

MORPHOLOGY: *Thallus* crustose to squamulose or foliose; upper surface corticate; cortex paraplectenchymatous of anticlinal hyphae; medulla woolly; lower surface ecorticate; attached to the substrate by a well developed dark prothallus or by rhizoids. *Ascocarp* an apothecium, immersed to adnate, laminal; disk round; margin proper; hypothecium pale to brown; paraphyses unbranched; ascus clavate, unitunicate, I+, I+ tholus, 8 spored; ascospores ellipsoid, simple hyaline, thin walled. *Pycnidia* immersed; fulcrum endobasidial; pycnospores bacilliform.
CHEMISTRY: (1 species) no secondary lichen substances.
PHYCOBIONT: *Nostoc.*
DISTRIBUTION: Queensland, New South Wales, Victoria and Tasmania. 10 species reported (50).

PARMELINA HALE: *Phytologia* 28, 481 (1974b)

MORPHOLOGY: *Thallus* foliose, lobed, radiate, loosely appressed to the substrate; lobes sub-linear to irregular, corticate; cortex of anticlinal hyphae forming paraplectechyma; lower cortex usually black with simple or squarrose rhizines to the margin; margins of lobes with simple cilia at least in the axils. *Ascocarp* an apothecium, laminal, sessile; disk concave, round; margin thalloid; hypothecium pale; paraphyses branched; ascus clavate, unitunicate, I+ blue, tholus I+ blue, 8 spored; ascospores ellipsoid, simple, hyaline, thin walled. *Pycnidia* immersed, fulcrum endobasidial, bayonet-like; pycnospores bacilliform.
CHEMISTRY: (46 species) atranorin (46), leucotylin (17), zeorin (17), secaleconic acid A (16), terpenes (15), gyrophoric acid (8), salazinic acid (7), galbinic acid (4), "horrescens" unknown (4), lecanoric acid (4), connorstictic acid (2), constictic acid (2), norstictic acid (2), fumarprotocetraric acid (1), lobaric acid (1), protocetraric acid (1), protolichesterinic acid (1), stictic acid (1).
PHYCOBIONT: *Trebouxia.*
DISTRIBUTION: In all states, especially on bark. 9 species reported (47).

PARMELIOPSIS (STIZ.) NYL.: *Syn. Lich.* 53 (1860)

MORPHOLOGY: *Thallus* foliose, appressed, irregularly lobed, upper and lower surfaces corticate; cortex of more or less anticlinal woven hyphae not forming paraplectenchyma; attached to the substrate by rhizines. *Ascocarp* an apothecium, laminal, sessile to adnate; disk round, concave; margin thalloid; hypothecium hyaline or brown; paraphyses rarely branched; ascus clavate, unitunicate, tholus I+ blue, 8 spored; ascospores ellipsoid to oblong, simple hyaline, thin walled. *Pycnidia* immersed; fulcrum exobasidial; pycnospores cylindrical, curved.
CHEMISTRY: (6 species) atranorin (5), divaricatic acid (3), thamnolic acid (2), usnic acid (2), protocetraric acid (1).
PHYCOBIONT: *Trebouxia.*
DISTRIBUTION: Victoria. 2 species reported (7).

PARMENTARIA FEE: *Essai Crypt.* 39 & 70 (1824)

MORPHOLOGY: Thallus crustose, uniform, epi- or endophloic, ecorticate or with an amorphous cortex. *Ascocarp* a perithecium; aggregated; the ostiolar canals oblique to horizontal and fusing in a common opening; naked or immersed in thalline or stromatic warts; paraphyses reticulate; ascus clavate, unitunicate, I−, 1-8 spored; ascospores ellipsoid to oblong, muriform, brown, walls unequally thickened resulting in lenticular or rhombic locules. *Pycnidia* immersed; fulcrum exobasidial.
CHEMISTRY: No data.
PHYCOBIONT: Trentepohlia.
DISTRIBUTION: Queensland and Victoria, in humid forests. 15 species reported (25).

PARMOTREMA MASS.: *Atti Ist. Veneto Sci. Ser 3(5),* 248 (1890)

MORPHOLOGY: Thallus foliose, lobed, radiate, loosely adnate to somewhat ascending, usually large; lobes broad, apically rotund with a distinct bare zone below the tips; upper and lower surfaces with a pallisade-plectenchymatous cortex, rhizines simple, often rather sparse. *Ascocarp* an apothecium; laminal, sub-stipitate to stipitate, often perforate, more or less round; disk concave, brown; margin thalloid; hypothecium pale; paraphyses branched; ascus clavate, unitunicate I+ blue, tholus I+ blue, 8 spored; ascospores ellipsoid, simple, hyaline, thin walled. *Pycnidia* immersed; fulcrum endobasidial, bayonet-like; pycnospores bacilliform.
CHEMISTRY: (106 species) atranorin (106), alectoronic acid (21), protocetraric acid (17), salazinic acid (17), protolichesterinic acid (14), usnic acid (12), rhodophyscin (10), gyrophoric acid (9), lecanoric acid (8), norstictic acid (7), cryptochlorophaeic acid (6), stictic acid (6), caperatic acid (3), olivetoric acid (2), vulpinic acid (2).
PHYCOBIONT: Trebouxia.
DISTRIBUTION: Widespread through humid and sub-humid areas, on bark, wood and rock. 26 species reported (110).

PECCANIA MASS.: *Atti Ist. veneto Sci. ser.* 3, 5, 335 (1860)

MORPHOLOGY: Thallus minutely fruticose of more or less branched lobes, forming small polsters; ecorticate; attached by rhizines. *Ascocarp* an apothecium, lateral to sub-terminal, constricted at the base; margin thalloid; hypothecium pale; paraphyses unbranched; asci clavate, unitunicate, 8-16 spored; ascospores ellipsoid, simple, hyaline, thin walled. *Pycnidia* immersed; fulcrum ?exobasidial; pycnospores ellipsoid to acicular.
CHEMISTRY: No data.
PHYCOBIONT: Xanthocapsa.
DISTRIBUTION: Western Australia, on calcareous substrates. ?1 species reported (14).

PELTIGERA WILLD.: *Flor. Berol. Prod.* 347 (1787)

MORPHOLOGY: Thallus foliose, spreading or somewhat ascending; upper and lower surfaces with a paraplectenchymatous cortex; upper surface with or without a tomentum; lower surface veined; attached to the substrate by bushy rhizoids. *Ascocarp* an apothecium, immersed on upper surface of specialized marginal lobes, emarginate; sometimes with a veil of cortical remains; hypothecium hyaline to brown; paraphyses unbranched; ascus cylindrical-clavate, unitunicate, I+ blue, tholus I+ blue, 8 spored, ascospores oblong-ellipsoid to acicular, transversely 3-7 septate, hyaline to brown, thin walled. *Pycnidia* immersed with filiform fulcra; pycnospores filiform.

CHEMISTRY: (22 species) tenuiorin (16), dolichorrhizin (8), zeorin (6), methylgyrophorate (5), scabrosin B (4), phlebic acid B (3), gyrophoric acid (2), phlebic acid A (2), scabrosin A (2), evernic acid (1), methylevernate (1), 4-0-methylgyrophoric acid (1), methyllecanorate (1), methylorsellinate (1), triterpenes (1), usnic acid (1).
PHYCOBIONT: *Nostoc* or *Coccomyxa.*
DISTRIBUTION: On soil, in humid areas of all states. 14 species reported (30).

PELTULA NYL.: *Annls. Sci. nat. Bot. ser 3,* 20, 316 (1853)

MORPHOLOGY: *Thallus* squamulose to peltate or sub-fruticose, corticate at least on the lower surface; paraplectenchymatous throughout or with a woolly medulla; attached to the substrate by a small area of rhizoids or an umbilicus. *Ascocarp* an apothecium, immersed, frequently covering the entire squamule; with or without a proper exciple; hypothecium pale; paraphyses unbranched; asci clavate, unitunicate, tholus I+ blue, many spored; ascospores minute, simple, hyaline, with a gelatinous sheath. *Pycnidia* immersed; fulcrum exobasidial; pycnospores bacilliform.
CHEMISTRY: (6 species) no determined secondary lichen substances.
PHYCOBIONT: *Anacystis.*
DISTRIBUTION: On soil and rock, especially in arid and semi-arid areas of Queensland, New South Wales, Victoria, South Australia and Western Australia. 7 species reported (18).

PERTUSARIA D.C. in LAM. et D.C.: *Flor. Franc.* 319 (1805)

MORPHOLOGY: *Thallus* crustose, uniform to dwarf fruticose, ecorticate or corticate with a paraplectenchymatous cortex of anticlinal hyphae; attached to the substrate by medullary or prothallial hyphae. *Ascocarp* an apothecium immersed in a wart, often perithecium-like in appearance, sometimes several hymenia in a wart, hymenia globose or disk-like, with profuse gelatin; hypothecium pale, paraphyses strongly reticulate; asci clavate, unitunicate (but with layers), thick walled, I+ blue, tholus I+ blue, the exoascus bursting in layers at maturity, 1-8 spored; ascospores ellipsoid or oblong-ellipsoid, simple, hyaline, walls often heavily thickened and ridged, commonly very large. *Pycnidia* immersed in warts; fulcrum exobasidial; pycnospores cylindrical to acicular or filiform.
CHEMISTRY: (84 species) thiophanic acid (23), norstictic acid (15), lichexanthone (14), stictic acid (14), picrolichenic acid (7), 2-7 dichloronorlichexanthone (5), fumarprotocetraric acid (4), atranorin (3), 2-7 dichlorolichexanthone (3), lecanoric acid (3), salazinic acid (3), thamnolic acid (3), gyrophoric acid (2), physodalic acid (2), arthothelin (1), connorstictic acid (1), haemathamnolic acid (1), 2-0-methylperlatolic acid (1), 4-0-methylperlatolic acid (1), perlatolic acid (1), protocetraric acid (1), stenosporic acid (1), variolaric acid (1).
PHYCOBIONT: *Trebouxia*-like.
DISTRIBUTION: On wood, bark and rocks, throughout the region. 65 species reported (250).

PHAEOGRAPHINA MULL. ARG.: *Flora, Jena* 65, 398 (1882)

MORPHOLOGY: *Thallus* crustose, uniform, epi- or endophloic, ecorticate or with a cortex of interwoven hyphae; attached by medullary or prothallial hyphae. *Ascocarp* an apothecium, elongate, often branched, adnate or immersed; disk narrow, proper exciple usually well developed; hypothecium pale to black; paraphyses unbranched; asci clavate, unitunicate, I−, usually 1-3 spored; ascospores oblong ellipsoid to elongate ellipsoid, large, muriform, brown,

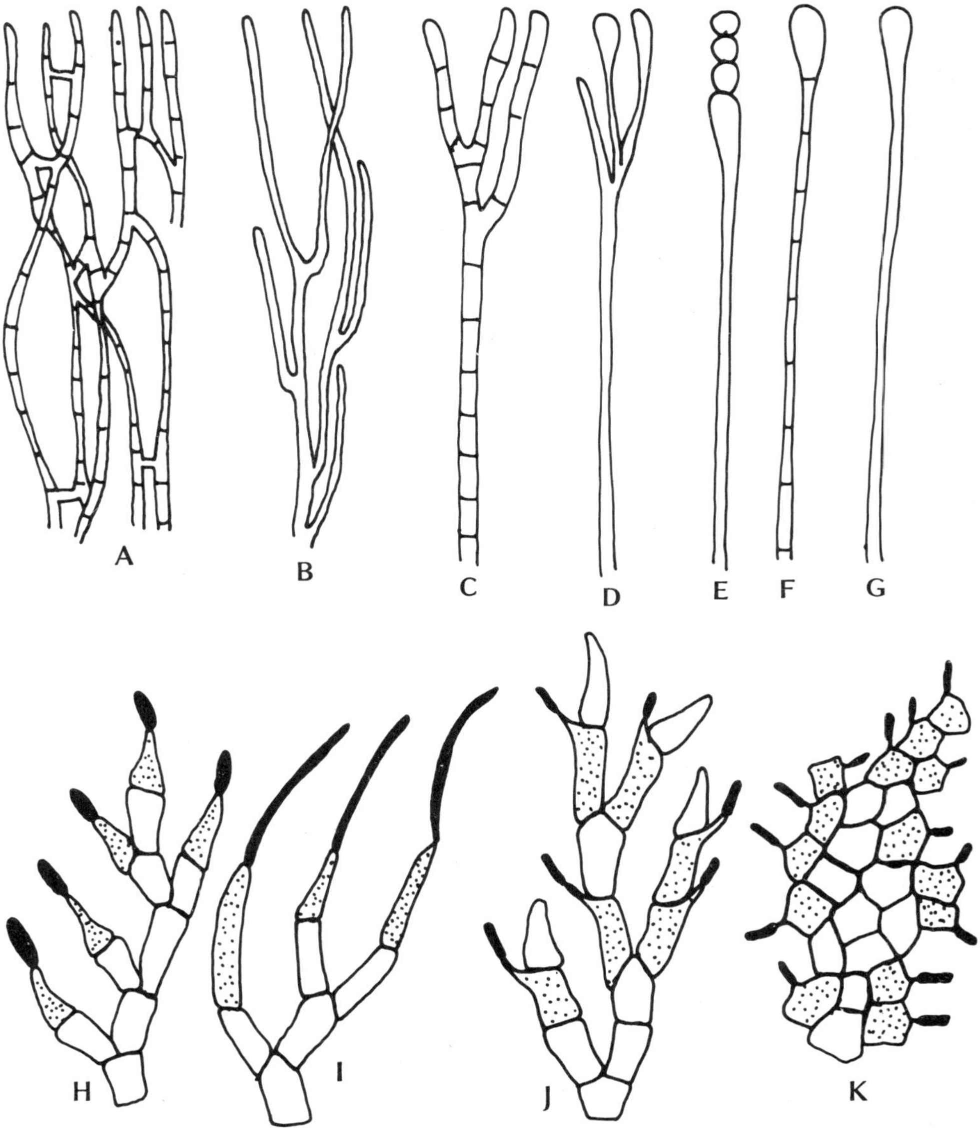

FIGURE 3 **Top:** Stylized variation in form of paraphyses and paraphysis-like structures: (A) reticulately branched, anastomosing, or net-like paraphyses, (B to D) variously branched paraphyses, (E to F) variations in unbranched paraphyses.
Bottom: Stylized pycnidial structures. (H and I) exobasidial fulcrum with pycnospores produced only by the terminal cells in a sequence, (H) shows ovoid pycnospores, (I) filiform pycnospores, (J and K) variants in the endobasidial form in which pycnospores are produced by cells other than terminal cells. (J) shows the bayonet-like form, and (K) the short celled form, both with bacilliform pycnospores. Pycnospores are shaded black, and the sterigmata producing them are stippled.

thin walled, I+ violet. *Pycnidia* rare, immersed; fulcrum exobasidial; pycnospores elongate, bacilliform.
CHEMISTRY: (7 species) norstictic acid (5), stictic acid (3).
PHYCOBIONT: Trentepohlia.
DISTRIBUTION: On wood or bark, in Queensland and Victoria. 5 species reported (90).

PHAEOGRAPHIS MULL. ARG.: *Flora, Jena* 65, 336 (1882)

MORPHOLOGY: Thallus crustose, uniform, epi- or endophloic; ecorticate or with a cortex of interwoven hyphae; attached by medullary or prothallial hyphae. *Ascocarp* an apothecium, immersed to sessile, elongate, often branched, disk narrow, proper exciple often black; thalloid exciple sometimes present; hypothecium pale to black; paraphyses unbranched; asci clavate, unitunicate, I− thick walled, 8 spored; ascospores transversely 3-many septate, brown, walls unequally thickened resulting in lenticular locules, I+ violet. *Pycnidia* rare, immersed; fulcrum exobasidial; pycnospores elongate or bacilliform.
CHEMISTRY: (4 species) norstictic acid (4).
PHYCOBIONT: Trentepohlia.
DISTRIBUTION: On wood or bark, in Queensland, New South Wales and Victoria. 15 species reported (200).

PHAEOPHYSCIA MOBERG: *Symb. bot. upsal.* 12(1), 29 (1877)

MORPHOLOGY: Thallus foliose, lobed, to more or less fruticose; loosely adnate; upper surface paraplectenchymatously corticate, grey to brown; lower surface paraplectenchymatously corticate, usually dark brown or black; attached to the substrate by simple rhizines. *Ascocarp* an apothecium, laminal, sessile, disk round; thalloid exciple well developed, usually with rhizines on the lower surface; disk black or brown; hypothecium hyaline; paraphyses unbranched or somewhat branched; asci cylindrical, unitunicate, I+ blue, tholus I+ blue, 8 spored; ascospores ellipsoid, 1 septate, brown, thick walled. *Pycnidia* immersed or partly immersed; fulcrum endobasidial; pycnospores minute ($<4\mu m$ long) ellipsoidal.
CHEMISTRY: (13 species) skyrin (10), zeorin (7), atranorin (6), leucotylin (6), emodin (1).
PHYCOBIONT: Trebouxia.
DISTRIBUTION: On bark, in Queensland, New South Wales and Victoria. 1 species reported (?13).

PHAEOTREMA MULL. ARG.: *Mem. Soc. Phys. Hist. nat. Geneve* 29(8), 10 (1887b)

MORPHOLOGY: Thallus crustose, uniform, granulose or warty, undifferentiated, or with a loose cortex; attached to the substrate by medullary or prothallial hyphae. *Ascocarp* an apothecium, more or less immersed in the thallus or in fruiting warts; disk circular or elongate, punctiform; the proper exciple initially almost closing the disk but eventually cracking and falling away, leaving fragments around the more or less free disk with a thalloid margin; hypothecium hyaline; paraphyses unbranched; ascus clavate to broadly clavate, unitunicate, I− (but hymenial gel may be I+ blue) thick walled, 1-8 spored; ascospores oblong to fusiform, transversely 3-many septate, brown or black, walls unequally thickened resulting in lenticular locules. *Pycnidia* not known.
CHEMISTRY: (8 species) 4-0-demethylnotatic acid (3), hypoprotocetraric acid (3), psoromic acid (2), conpsoromic acid (1), constictic acid (1), stictic acid (1).

PHYCOBIONT: *Trentepohlia.*
DISTRIBUTION: On bark, in humid forests of Queensland. 5 species recorded (30).

PHLYCTELLA KREMP.: *Verh. zool.-bot. Ges. Wein* 26, 452 (1876)

MORPHOLOGY: *Thallus* crustose, uniform, chinky or powdery, ecorticate, attached to the substrate by medullary hyphae. *Ascocarp* an apothecium, immersed to adnate; disk round, concave to flat; thalloid margin often poorly developed, irregular; hypothecium hyaline to brown; paraphyses unbranched; asci clavate, unitunicate, I?, 1-8 spored; ascospores fusiform, transversely 3-many septate, walls unequally thickened to produce lenticular locules. *Pycnidia* immersed; fulcrum exobasidial; pycnospores oblong, straight.
CHEMISTRY: No data.
PHYCOBIONT: *Trebouxia*-like.
DISTRIBUTION: On bark, in Victoria. 1 species reported (15).

PHYLLOPSORA MULL. ARG.: *Bull. Herb. Boissier* 2, app. 1, 11 & 45 (1894a)

MORPHOLOGY: *Thallus* minutely squamulose to much dissected foliose, upper surface corticate; attached to the substrate by a tomentum of rhizoids. *Ascocarp* an apothecium, sessile; disk round; margin proper; hypothecium pale to dark; paraphyses unbranched; asci cylindrical, unitunicate, I+, tholus I+blue, thin walled, 8 spored; ascospores oblong-ellipsoid to acicular, simple, hyaline, thin walled. *Pycnidia* immersed; fulcrum exobasidial; pycnospores cylindrical, straight.
CHEMISTRY: No data.
PHYCOBIONT: *Trebouxia*-like.
DISTRIBUTION: In humid forests of Queensland, Victoria and Tasmania. 4 species reported (35).

PHYSCIA (SCHREB.) TH. FR. em VAIN.: *Acta Soc. Fauna Flora fenn.* 7(1), 138 (1890)

MORPHOLOGY: *Thallus* foliose, lobed, loosely adnate; upper surface paraplectenchymatously corticate, pale to dark grey; lower cortex prosoplectenchymatous, white to pale brown; attached to the substrate by simple rhizines. *Ascocarp* an apothecium, laminal, sessile, disk round; thalloid exciple well developed, with rhizines on the lower surface; disk brown or black; hypothecium hyaline; paraphyses unbranched or somewhat branched; asci cylindrical, unitunicate, I+ blue, tholus I+ blue, 8 spored; ascospores ellipsoid, 1 septate, brown, thick walled towards the apices. *Pycnidia* immersed; fulcrum endobasidial; pycnospores cylindrical, short ($>4\mu m$ $<10\mu m$ long).
CHEMISTRY: (13 species) atranorin (13), leucotylin (11), triterpenoids (11), zeorin (11).
PHYCOBIONT: *Trebouxia.*
DISTRIBUTION: Widespread on trees and rocks, especially in the southern half of Australia. 17 species reported (?120).

PHYSCIOPSIS CHOISY: *Bull. mens. Soc. linn. Lyon* 19, 20 (1950)
(*Physciopsis* Choisy is a synonym of *Hyperphyscia* Mull. Arg. p.10, (1894b)
MORPHOLOGY: *Thallus* foliose, lobed, closely adnate; upper surface paraplectenchymatously corticate, olive or brown; lower cortex prosoplectenchymatous if present, brown, often hard to detect; attached either directly or by very short indistinctly

delimited rhizines. *Ascocarp* an apothecium, laminal, sessile; disk round, brown to black; margin thalloid, without rhizines on the lower surface; hypothecium hyaline; paraphyses unbranched or somewhat branched; asci cylindrical, unitunicate, I+ blue, tholus I+ blue, 8 spored; ascospores ellipsoid, 1(-3?) transversely septate, brown, walls thickened toward the apices. *Pycnidia* immersed, fulcrum endobasidial; pycnospores filiform (>10μm long).
CHEMISTRY: (3 species) no secondary lichen substances known.
PHYCOBIONT: Trebouxia.
DISTRIBUTION: Throughout the region, but usually obscure. 2 species reported (10-15).

PHYSMA MASS.: *Neag. Lich.* 6 (1854a)

MORPHOLOGY: Thallus foliose, lobed, radiate, corticate; cortex paraplectenchymatous; attached to the substrate directly or by rhizines forming a tomentum. *Ascocarp* an apothecium, sessile or sub-stipitate; disk round, concave, margin thalline; hypothecium hyaline; paraphyses simple; ascus clavate, unitunicate, tholus I+, ascal gelatin I+ blue, 8 spored; ascospores ellipsoid to fusiform, simple, hyaline, thick walled, sometimes minutely verrucose. *Pycnidia* immersed; fulcrum exobasidial; pycnospores short, cylindrical or constricted in the centre.
CHEMISTRY: No data.
PHYCOBIONT: Nostoc.
DISTRIBUTION: On bark, in Queensland, New South Wales and Victoria. 4 species reported (10).

PILOPHORUS TH.FR.: *De Stereocaulis et Pilophoris Commentatio* 40 (1857)

MORPHOLOGY: Thallus granular crustose, warty to squamulose, ecorticate; undifferentiated; attached to the substrate by medullary hyphae; bearing well developed podetia; podetia upright, rarely branched, cylindrical, ecorticate, often granular-squamulose; sometimes with cephalodia; outer medulla loose and containing algae, inner medulla denser, of longitudinal hyphae, solid or hollow. *Ascocarp* an apothecium, apothecia single or aggregate, capitate or semi-globose; proper exciple dissappearing early; hypothecium dark; paraphyses unbranched; ascus narrow-clavate, unitunicate, I+ blue, tholus I+ blue, 8 spored; ascospores ellipsoid to oblong-ellipsoid, simple, hyaline, thin walled. *Pycnidia* adnate on tips of podetia; fulcrum exobasidial; pycnospores cylindrical or curved.
CHEMISTRY: (10 species) atranorin (9), zeorin (5), stictic acid (2), caperatic acid (1), constictic acid (1).
PHYCOBIONT: Trebouxia, cephalodia contain *Stigonema.*
DISTRIBUTION: New South Wales and Victoria. 2 species reported (10).

PLACOPSIS (NYL.) LINDSAY: *Trans. Linn. Soc. London* 25, 536 (1866)

MORPHOLOGY: Thallus crustose, with marginal lobes, corticate; cortex paraplectenchymatous; primary phycobiont commonly more or less displaced by internal cephalodia, with developed, apothecium-like external cephalodia. *Ascocarp* an apothecium, sessile; disk round, margin thalloid; hypothecium hyaline; paraphyses slender, branched, moniliform near the tip; ascus cylindrical, unitunicate, ?I−, 8 spored; ascospores ellipsoid, simple, hyaline, thin walled. *Pycnidia* immersed; fulcrum exobasidial; pycnospores bacilliform.
CHEMISTRY: (12 species) gyrophoric acid (12), lecanoric acid (9), phenolics (1).
PHYCOBIONT: Initially *Trebouxia*-like, with internal and external *Stigonema* cephalodia.
DISTRIBUTION: Queensland, New South Wales, Victoria and Tasmania, on rocks or soil. 5 species reported (34).

PLACYNTHIUM (ACH.) GRAY: *Nat. Arr. Br. Pl.* 1, 395 (1821)

MORPHOLOGY: *Thallus* crustose or squamulose to foliose, irregularly lobed, sometimes with a developed hypothallus; well developed paraplectenchymatous upper cortex; poorly developed lower cortex; medulla more or less paraplectenchymatous. *Ascocarp* an apothecium, sessile; proper margin thin; hypothecium hyaline or dark; paraphyses thick, unbranched; ascus clavate, unitunicate, tholus I+, 8 spored; ascospores elongate or ellipsoid, transversely 2-8 septate, hyaline, thin walled. *Pycnidia* immersed; fulcrum endobasidial; pycnospores bacilliform.
CHEMISTRY: No data.
PHYCOBIONT: *Dichothrix.*
DISTRIBUTION: New South Wales. 2 species reported (25).

PLEUROTHELIOPSIS ZAHLBR.: *Cat. Lich. Univ.* 1, 512 (1922)

MORPHOLOGY: *Thallus* crustose, uniform; epi- or endophloic; ecorticate or with a cortex of periclinal hyphae. *Ascocarp* a perithecium opening by a more or less oblique ostiolar canal, usually clad in thalline material; paraphyses reticulate; asci cylindrical-clavate, bitunicate, usually thick walled, I−, 1-8 spored; ascospores muriform, brown, thin walled, locules angular. *Pycnidia* not known.
CHEMISTRY: No data.
PHYCOBIONT: *Trentepohlia.*
DISTRIBUTION: Queensland, on bark. 1 species reported (6).

PLEUROTREMA MULL. ARG.: *Bot. Jb.* 6, 388 (1885)

MORPHOLOGY: *Thallus* crustose, uniform, epi- or endophloic; ecorticate. *Ascocarp* a perithecium, perithecia sessile or immersed; simple or compound; ostiolar canals oblique or prostrate; paraphyses reticulate; ascus cylindrical, bitunicate, thick walled, I−, 8 spored; ascospores oblong-ellipsoid to filiform, transversely 1-many septate, hyaline, thin walled. *Pycnidia* not known.
CHEMISTRY: No data.
PHYCOBIONT: *Trentepohlia.*
DISTRIBUTION: On bark, in Queensland. 1 species reported (10).

POLYBLASTIOPSIS ZAHLBR.: *Pflanzenfamilien* 1, 67 (1903)

MORPHOLOGY: *Thallus* crustose, uniform, endo- or epiphloic, ecorticate. *Ascocarp* a perithecium-like pseudothecium or perithecium, immersed or sessile, with an erect ostiole; paraphyses reticulate; ascus bitunicate, thick walled, I−, 1-8 spored; ascospores oval to oblong, muriform, hyaline, thin walled. *Pycnidia* not known.
CHEMISTRY: No data.
PHYCOBIONT: *Trochisia* and other genera.
DISTRIBUTION: On bark or soil, in Queensland. 7 species reported (?18).

POLYCAULONIA HUE: *Exp. Antarct. Franc. Sci. Nat. Lich.* 8 (1908)

MORPHOLOGY: *Thallus* fruticose, erect or decumbent, dichotomous or irregularly branched, fully corticate, yellow. *Ascocarp* an apothecium, terminal, constricted at the base;

disk round, concave, orange; margin thalloid; hypothecium hyaline; paraphyses unbranched; asci cylindrical, unitunicate, I+ blue, tholus I+ blue, 8 spored; ascospores ellipsoid, polaribilocular, hyaline. *Pycnidia* terminal on axes; fulcrum endobasidial; pycnospores bacilliform.
CHEMISTRY: (1 species) parietin.
PHYCOBIONT: *Trebouxia.*
DISTRIBUTION: Tasmania. ?1 species reported (20).

PORINA MULL. ARG.: *Flora, Jena* 66, 320 (1883)

MORPHOLOGY: *Thallus* crustose, uniform, epi- or endophloic, ecorticate. *Ascocarp* a perithecium, solitary, immersed or sessile; ostiole erect; paraphyses unbranched or little branched; ascus clavate, bitunicate, thin walled, I−, 8 spored; ascospores ellipsoid to acicular, transversely 1-many septate, hyaline, thin walled. *Pycnidia* immersed, fulcrum exobasidial; pycnospores bacilliform to filiform.
CHEMISTRY: No data.
PHYCOBIONT: *Trentepohlia, Phycopeltis.*
DISTRIBUTION: Queensland, New South Wales and Victoria in humid regions. 38 species reported (300).

POROCYPHUS KORB.: *Syst. Lich. Germ.* 425 (1855)

MORPHOLOGY: *Thallus* crustose, smooth to granular or minutely fruticose; attached by medullary hyphae; ecorticate; gelatinous. *Ascocarp* an apothecium, immersed, more or less closed by the thalloid margin, proper exciple pale and narrow; hypothecium hyaline; paraphyses fine, unbranched; ascus clavate, unitunicate, thin walled, 8 spored; ascospores ovoid to ellipsoid, simple, hyaline, thin walled. *Pycnidia* immersed; fulcrum exobasidial; pycnospores short, oblong-ellipsoid.
CHEMISTRY: No data.
PHYCOBIONT: *Calothrix.*
DISTRIBUTION: On rocks, in Western Australia. 1 species reported (7).

PROTOBLASTENIA STEINER: *Verh. zool.-bot. Ges. Wein* 61, 47 (1911)

MORPHOLOGY: *Thallus* crustose, uniform, granular, yellow, ecorticate, attached to the substrate by medullary hyphae. *Ascocarp* an apothecium, immersed or sessile; disk round, yellow; margin proper; hypothecium pale or dark; paraphyses unbranched; ascus clavate, unitunicate, thin walled, I+ blue, tholus I+ blue, 8 spored; ascospores ellipsoid, simple, hyaline, thin walled. *Pycnidia* immersed; fulcrum endobasidial; pycnospores bacilliform.
CHEMISTRY: (8 species) emodin (8), parietin (7), fallacinal (6), parietinic acid (6), teloschistin (6), erythroglaucin (5), xanthorin (4), usnic acid (1).
PHYCOBIONT: *Trebouxia.*
DISTRIBUTION: On rocks and bark, probably in all states. ?3 species reported (11).

PSEUDEVERNIA ZOPF.: *Bei. Bot. Central.* 14, 124 (1903)

MORPHOLOGY: *Thallus* foliose, prostrate or ascending; lobes markedly recurved and channelled below; upper and lower surfaces paraplectenchymatous; lower surface often

becoming black, devoid of rhizines; lobes eciliate. *Ascocarp* an apothecium, stipitate; hypothecium pale; paraphyses unbranched or little branched; asci cylindrical, unitunicate, ?I+ blue, ?tholus I+ blue, 8 spored; spores ellipsoid, simple, hyaline. *Pycnidia* immersed, fulcrum ?endobasidial; pycnospores ?bacilliform.
CHEMISTRY: (5 species) atranorin (5), physodic acid (3), chloroatranorin (2), lecanoric acid (2), olivetoric acid (1), tetrahydroxy fatty acids (1).
PHYCOBIONT: *Trebouxia.*
DISTRIBUTION: Tasmania. 1 species reported (5).

PSEUDOCYPHELLARIA VAIN.: *Acta Soc. Fauna Flora fenn.* 7(1), 182 (1890)

MORPHOLOGY: *Thallus* foliose, lobed, radiate; both surfaces with a paraplectenchymatous cortex; medulla loose; attached to the substrate by rhizoids which often form a mat; lower surface broken by pseudocyphellae. *Ascocarp* an apothecium, sub-stipitate, marginal or laminal; disk round, usually concave; thalloid exciple; hypothecium pale or dark; paraphyses unbranched; ascus clavate, unitunicate, I+ blue, 8 spored; ascospores oblong-fusiform to fusiform, transversely 1-several septate, hyaline becoming brown, thin walled. *Pycnidia* immersed; fulcrum endobasidial; pycnospores short, straight, cylindrical.
CHEMISTRY: (44 species) calycin (21), pulvinic dilactone (21), tenuiorin (20), 7β-acetoxy-22-hydroxyhopane (19), constictic acid (16), stictic acid (16), 15α-22-dihydroxyhopane (12), norstictic acid (9), pulvinic acid (9), zeorin (7), durvillonol (5), durvilldiol (4), methylgyrophorate (3), gyrophoric acid (2), 4-0-methylgyrophoric acid (2), salazinic acid (2), atranorin (1) ergosterol (1), n-heptadecane (1), 4-0-methylbarbatic acid (1), methylevernate (1), methyllecanorate (1), methylorsellinate (1), 4-0-methylorsellinic acid (1), orsellinic acid (1), pannarin (1), protocetraric acid (1), usnic acid (1).
PHYCOBIONT: *Nostoc, Palmella* or *Trebouxia*-like.
DISTRIBUTION: In humid parts of Australia, on rock and bark. 43 species reported (200).

PSEUDOPARMELIA LYNGE: *Ark. Bot.* 13(13), 15 (1914)

MORPHOLOGY: *Thallus* foliose, lobed, radiate, adnate to the substrate; lobes linear irregular, corticate; cortex of anticlinal hyphae forming paraplectenchyma; lower cortex black to pale brown with simple rhizines to the margin; margins sometimes black rimmed, eciliate. *Ascocarp* an apothecium, laminal, sessile; disk concave, round; margin thalloid; hypothecium pale; paraphyses branched; ascus clavate, unitunicate, I+ blue, tholus I+ blue, 8 spored; ascospores ellipsoid, simple, hyaline, thin walled. *Pycnidia* immersed; fulcrum endobasidial, bayonet-like; pycnospores bacilliform.
CHEMISTRY: (76 species) atranorin (55), protocetraric acid (25), usnic acid (20), divaricatic acid (9), stictic acid (8), salazinic acid (7), caperatic acid (5), constictic acid (5), gyrophoric acid (4), lecanoric acid (4), norlobaridone (3), protolichesterinic acid (3), neoloxodic acid (2), norstictic acid (2), perlatolic acid (2), physodalic acid (2), skyrin (2), barbatic acid (1), cryptochlorophaeic acid (1), 4-0-demethylbarbatic acid (1), evernic acid (1), fumarprotocetraric acid (1), obtusatic acid (1), sekikaic acid (1), succinprotocetraric acid (1).
PHYCOBIONT: *Trebouxia.*
DISTRIBUTION: Throughout Australia, especially on trees. 19 species reported (76).

PSEUDOPYRENULA MULL. ARG.: *Flora, Jena,* 66, 247 (1883)

MORPHOLOGY: *Thallus* crustose, uniform, endo- or epiphloic, undifferentiated, ecorticate. *Ascocarp* a perithecium, perithecia scattered or grouped, either naked or immersed; ostiole

erect; paraphyses reticulate; ascus clavate to cylindrical, bitunicate, I−, 8 spored; ascospores oblong to oblong-ellipsoid, transversely 3 or more septate, hyaline, walls unevenly thickened resulting in lenticular or angular locules. *Pycnidia* immersed; fulcrum exobasidial, pycnospores bacilliform, constricted in the middle.
CHEMISTRY: No data.
PHYCOBIONT: *Trentepohlia.*
DISTRIBUTION: Queensland and Tasmania. 3 species reported (45).

PSORA HOFFM.: *Descr. adumb Lich.* 1, 37, (1790)

MORPHOLOGY: *Thallus* squamulose, corticate, attached to the substrate by medullary hyphae. Ascocarp an apothecium, sessile, disk round; margin proper; hypothecium pale or coloured, paraphyses branched or unbranched; asci cylindrical-clavate, unitunicate, tholus I+ blue, 8-spored; ascospores ellipsoid, simple, hyaline, thin-walled. *Pycnidia* immersed; fulcrum exobasidial; pycnospores bacilliform.
CHEMISTRY: (3 species) lecanoric acid (1), norstictic acid (1), rhizocarpic acid (1).
PHYCOBIONT: *Trebouxia*-like.
DISTRIBUTION: On soil especially in drier areas. 7 species reported (50).

PSOROMA NYL.: *Mem. Soc. Sci. Nat. Cherbourg* 2, 5-16 (1855)

MORPHOLOGY: *Thallus* squamulose to foliose, more or less irregular; developed paraplectenchymatous upper cortex of anticlinal hyphae and a thin lower cortex of interwoven periclinal hyphae; attached to the substrate by a few rhizoids. *Ascocarp* an apothecium, laminal, adnate to sessile; thalloid margin; hypothecium hyaline; paraphyses unbranched; asci clavate, unitunicate, tholus I+ blue, 8 spored; ascospores ellipsoid to spherical, simple, hyaline, thin walled. *Pycnidia* immersed; fulcrum endobasidial; pycnospores short, cylindrical or constricted in the middle.
CHEMISTRY: (2 species) pannarin (1), zeorin (1).
PHYCOBIONT: *Myrmecia.*
DISTRIBUTION: In all humid parts of the region, on rock, soil and bark. 16 species reported (35).

PSOROMARIA NYL.: *Lich. N.Z.* 54 (1888)

MORPHOLOGY: *Thallus* squamulose to foliose, more or less irregular; developed paraplectenchymatous upper cortex of anticlinal hyphae and a thin lower cortex of interwoven periclinal hyphae; attached to the substrate by a few rhizoids. *Ascocarp* an apothecium, laminal, adnate to sessile; proper margin; hypothecium hyaline; paraphyses unbranched; asci clavate, unitunicate, tholus I+ blue, 8 spored; ascospores ellipsoid to spherical, simple, hyaline, thin walled. *Pycnidia* immersed; fulcrum endobasidial; pycnospores short, cylindrical or constricted in the middle.
CHEMISTRY: No data.
PHYCOBIONT: *Myrmecia.*
DISTRIBUTION: Tasmania. ?1 species reported (2?)

PYRENOPSIDIUM (NYL.) FORSS.: *Nova Acta R. Soc. Scient. Upsal. Ser 3,* 13, 39 & 59 (1885)

MORPHOLOGY: Thallus crustose, uniform, granular or areolate, gelatinous, ecorticate; of loose, much branched hyphae enmeshing algal cells; attached to the substrate by medullary hyphae. *Ascocarp* an apothecium, immersed to sessile, initially closed, later open, margin thalloid; disk often narrow; hypothecium hyaline to brown; paraphyses unbranched; ascus clavate, unitunicate, ?I+, 8 spored; ascospores oblong to globose, simple, hyaline, thin walled. *Pycnidia* immersed; fulcrum exobasidial; pycnospores ellipsoid-oblong.
CHEMISTRY: No data.
PHYCOBIONT: Chroococcus.
DISTRIBUTION: On rock, in arid Western Australia. 1 species reported (8).

PYRENOPSIS NYL.: *Syn. Lich.* 1, 97 (1858)

MORPHOLOGY: Thallus crustose, uniform, granular to minutely fruticose, gelatinous, ecorticate, of hyphae enmeshing the phycobiont. *Ascocarp* an apothecium, immersed to sessile, initially closed, later open; thalloid exciple well developed; hypothecium pale to brown; paraphyses unbranched, often indistinct; ascus clavate, unitunicate, ?I+, 8-32 spored; ascospores oblong to spherical, simple, hyaline, thin walled. *Pycnidia* immersed; fulcrum exobasidial, bushy; pycnospores oblong or rarely filiform and curved.
CHEMISTRY: No data.
PHYCOBIONT: Gloeocapsa.
DISTRIBUTION: Victoria and Tasmania. 2 species reported (40).

PYRENULA MASS.: *Ric. Lich. Crost.,* 162 (1852)

MORPHOLOGY: Thallus crustose, uniform, endo- or epiphloic; ecorticate. *Ascocarp* a perithecium, immersed or naked; ostiole erect; paraphyses fine, unbranched; ascus clavate, bitunicate, I−, 8 spored; ascospores ellipsoid to fusiform, transversely 3-7 septate, brown, walls unevenly thickened resulting in lenticular or rhomboidal locules. *Pycnidia* immersed; fulcrum exobasidial; pycnospores filiform, curved.
CHEMISTRY: (3 species) anthraquinoid (1), lichexanthone (1), parietin (1), phenolic (1).
PHYCOBIONT: Trentepohlia.
DISTRIBUTION: Queensland, New South Wales, Victoria and Tasmania, in humid regions. 35 species reported (190).

PYRGILLUS NYL.: *Syn. Lich.* 1, 168 (1860)

MORPHOLOGY: Thallus crustose, uniform, ecorticate. *Ascocarp* an apothecium, immersed to adnate, urn-like, disk narrow and flat; hypothecium pale; paraphyses thread-like; ascus thin walled, unitunicate, disintegrating into a mazaedium, 8 spored; spores oblong to elliptical, 1-3 septate, brown, walls unequally thickened to produce lenticular locules. *Pycnidia* immersed; fulcrum exobasidial; pycnospores filiform, curved.
CHEMISTRY: No data.
PHYCOBIONT: Trentepohlia.
DISTRIBUTION: Queensland, in humid forests. 4 species reported (8).

PYXINE FR.: *Syst. orb.* 1, 267 (1825)

MORPHOLOGY: Thallus foliose, appressed, radiate, both surfaces corticate; the upper surface paraplectenchymatous of anticlinal hyphae, the lower of longitudinal hyphae; attached to the substrate by rhizines. *Ascocarp* an apothecium, sessile, laminal; margin sometimes with thalloid rim disappearing or retained, usually proper; hypothecium dark; paraphyses unbranched; ascus clavate, unitunicate, I+ blue, tholus I+ blue, 8 spored; ascospores oblong to ellipsoid, 1 septate, brown, walls thickened, lumina small. *Pycnidia* immersed; fulcrum endobasidial; pycnospores short, cylindrical.
CHEMISTRY: (32 species) unknown terpenoids (20), atranorin (16), lichexanthone (11), norstictic acid (7), dammarane derivatives (2), methylpyxinate (2), pyxinic acid (2), chiodectonic acid (1), chloroatranorin (1), diacetyl pyxinol (1), pyxinol (1), skyrin (1).
PHYCOBIONT: Trebouxia.
DISTRIBUTION: Queensland and New South Wales, especially near the coast. 11 species reported (35).

RAMALEA NYL.: *Flora, Jena* 49, 289 (1866)

MORPHOLOGY: Thallus squamulose to dwarfed fruticose, partly horizontal, tufted; both surfaces corticate; without rhizines; attached by hyphae. *Ascocarp* an apothecium, circular, lateral, stipitate; margin proper, hypothecium pale; paraphyses unbranched; ascus clavate, unitunicate, ?I+, 8 spored; ascospores acicular, simple or rarely 1 septate. *Pycnidia* immersed in the thallus; fulcrum exobasidial; pycnospores cylindrical, curved.
CHEMISTRY: (1 species) usnic acid (1).
PHYCOBIONT: Trebouxia.
DISTRIBUTION: On acid soils in heathlands, in Western Australia and Victoria. 1 species reported (3).

RAMALINA ACH.: *Lich. Univ.* 122 (1810)

MORPHOLOGY: Thallus fruticose, erect or pendulous, all surfaces corticate of anticlinal hyphae; strengthened by chondroid fibres of periclinal hyphae which may form a continuous ring; medulla sparse; attached to the substrate by a basal disk; commonly bilaterally flattened, sometimes dorsiventrally differentiated. *Ascocarp* an apothecium, terminal, sub-terminal, marginal or laminal; sessile or stipitate; concave; thalloid exciple; hypothecium pale; paraphyses unbranched; ascus clavate, unitunicate, I+ blue, 8 spored; ascospores ellipsoid to fusiform, often curved, 1-3 septate, hyaline, thin walled. *Pycnidia* wholly or partly immersed in the thallus; fulcrum exobasidial, little branched; pycnospores bacilliform, straight.
CHEMISTRY: (85 species) usnic acid (74), salazinic acid (19), obtusatic acid (14), atranorin (11), norstictic acid (10), divaricatic acid (9), sekikaic acid (9), protocetraric acid (6), evernic acid (5), stictic acid (4), hypoprotocetraric acid (3), ramalinolic acid (3), chloroatranorin (2), stenosporic acid (2), boninic acid (4), bourgeanic acid (1), constictic acid (1), cryptochlorophaeic acid (1), 4-0-demethylbarbatic acid (1), 1-methyl-3,5 dichlorolecanorate (1), paludosic acid (1), perlatolic acid (1).
PHYCOBIONT: Trebouxia.
DISTRIBUTION: On bark and rocks, in all but the most arid parts of Australia. 41 species reported (200).

RELICINA HALE: *Phytologia* 28, 484 (1974b)

MORPHOLOGY: Thallus foliose; attached to the substrate by rhizines; with well developed upper and lower cortices, upper cortex pallisade plectenchyma; lower surface brown to black, rhizines simple or sparsely branched to squarrose or bushy and forming a tomentum; margins of lobes with bulbate cilia. *Ascocarp* an apothecium; sessile; margin thalloid, with or without a corona of sessile pycnidia, with or without bulbate cilia on the lower surface; disk round, flat or concave; hypothecium pale; paraphyses unbranched, capitate; asci cylindrical-clavate, unitunicate I+ blue, 8 spored; ascospores simple, ovoid to bicornute, thick walled. *Pycnidia* immersed in the thallus or sessile on apothecial margins; fulcrum endobasidial, bayonet-like; pycnospores bacilliform.
CHEMISTRY: (24 species) usnic acid (24), echinocarpic acid (6), atranorin (5), fumarprotocetraric acid (5), norstictic acid (4), salazinic acid (4), barbatic acid (3), 4-0-methyl barbatic acid (3), protocetraric acid (3), succinprotocetraric acid (3), diffractaic acid (2), stictic acid (2), alectoronic acid (1), caperatic acid (1), consalazinic acid (1), gyrophoric acid (1), "quintaria" unknown (1), protolichesterinic acid (1).
PHYCOBIONT: Trebouxia.
DISTRIBUTION: Queensland and New South Wales, on rocks and on bark in humid forests. 8 species reported (24).

RHIZOCARPON LAM. ex LAM & DC.: *Flor. Franc.* 365 (1805)

MORPHOLOGY: Thallus crustose, uniform; ecorticate; often with a well developed prothallus; attached to the substrate by prothallial hyphae. *Ascocarp* an apothecium; immersed to sessile; disk more or less round, black; margin proper, black, disappearing; hypothecium dark; paraphyses usually reticulate; ascus clavate or inflated-clavate, unitunicate, I+, tholus I+ blue, 1-8 spored; ascospores irregular, muriform, brown, mucilagenous. *Pycnidia* immersed; fulcrum exobasidial; pycnospores cylindrical to acicular, straight or curved.
CHEMISTRY: (32 species) stictic acid (18), rhizocarpic acid (16), psoromic acid (11), gyrophoric acid (7), barbatic acid (4), norstictic acid (4), lecanoric acid (1), physodic acid (1).
PHYCOBIONT: Trebouxia.
DISTRIBUTION: Queensland, New South Wales, Victoria, Tasmania and South Australia, on rocks in a wide range of habitats. 7 species reported (200).

RINODINA (ACH.) GRAY: *Nat. Arr. Br. Pl.* 448 (1821)

MORPHOLOGY: Thallus crustose to rarely squamulose, uniform or marginally lobed; ecorticate or with a paraplectenchymatous upper cortex; attached to the substrate by medullary or prothallial hyphae. *Ascocarp* an apothecium, immersed to sessile; margin initially with a thalloid covering, later proper or emarginate; disk dark; hypothecium hyaline; paraphyses rarely branched; ascus clavate, unitunicate, thick walled, I+ blue, tholus I+ blue, usually 8 but up to 24 spored; ascospores transversely 1-3 septate, grey or brown, walls unevenly thickened resulting in small locules linked by an isthmus. *Pycnidia* immersed or prominent in warts; fulcrum endobasidial; pycnospores bacilliform.
CHEMISTRY: (23 species) atranorin (15), norstictic acid (4), chloroatranorin (2), gyrophoric acid (2), lecanoric acid (2), zeorin (2), α-collatolic acid (1), fumarprotocetraric acid (1), phenolics (1), usnic acid (1).
PHYCOBIONT: Trebouxia.
DISTRIBUTION: Queensland, Victoria, Tasmania, South Australia and Western Australia, on rock and bark. 14 species reported (200).

ROCELLA DC. apud LAM. & DC.: *Flor. Franc* ed 3, 2, 334 (1805)

MORPHOLOGY: Thallus fruticose, upright or pendant, branched, flat or terete; cortex of transverse hyphae; attached by a basal disk. *Ascocarp* an apothecium-like pseudothecium, round or somewhat irregular, sessile, lateral; margin thalloid or with a proper exciple; hypothecium black; paraphyses branched; ascus clavate, bitunicate, I−, 8 spored; ascospores oblong to fusiform, transversely several septate, hyaline, thin walled. *Pycnidia* immersed, lateral; fulcrum exobasidial; pycnospores bacilliform, curved.
CHEMISTRY: (37 species) erythrin (23), lecanoric acid (19), rocellic acid (9), mesoerythritol (6), portentol (3), acetylportentol (2), erythritol (2), lepraric acid (2), norstictic acid (2), tetrahydroxy fatty acids (2), aspicilin (1), ethyl orsellinate (1), galapagin (1), montagnetol (1), orcinol (1), picroerythrin (1), protocetraric acid (1), skyrin (1), zeorin (1).
PHYCOBIONT: Trentepohlia.
DISTRIBUTION: On mangroves, in the tropics. 2 species reported (35).

SAGENIDIUM STIRT.: *Proc. Phil. Soc. Glasgow* 10, 305 (1877)

MORPHOLOGY: Thallus fruticose, branched, flat or terete, corticate, attached by a basal disk. *Ascocarp* an apothecium-like pseudothecium, sessile; without a margin; hypothecium black; paraphyses branched; ascus clavate, bitunicate, I−, 8 spored; ascospores hyaline, transversely several septate. *Pycnidia* not known.
CHEMISTRY: (4 species) atranorin (1), lecanoric acid (1), lepraric acid (1), protocetraric acid (1), schizopeltic acid (1).
PHYCOBIONT: Chlorophycean.
DISTRIBUTION: Tasmania, on ?maritime rocks. 1 species reported (4).

SAGIOLECHIA MASS.: *Geneac. Lich.* 11 (1854b)

MORPHOLOGY: Thallus crustose, uniform, ecorticate; attached to the substrate by medullary hyphae. *Ascocarp* an apothecium, initially immersed, eventually sessile, round; proper margin thin, dark; thalloid margin sometimes disappearing; hypothecium dark; paraphyses simple; ascus clavate, unitunicate, I−, 8 spored; ascospores ellipsoid to fusiform, transversely 2-3 septate, hyaline, uniformly thin walled. *Pycnidia* immersed; fulcrum exobasidial; pycnospores short.
CHEMISTRY: No data.
PHYCOBIONT: Trentepohlia.
DISTRIBUTION: Queensland, on calcareous rock. 1 species reported (2).

SARCOGRAPHA FEE: *Essai Crypt.* 35 & 58 (1824)

MORPHOLOGY: Thallus crustose, uniform; ecorticate or with a poorly developed cortex of periclinal hyphae; attached by medullary or prothallial hyphae. *Ascocarp* an apothecium, elongate or irregular; immersed in stroma which may be poorly developed; well developed proper exciple; hypothecium pale; hymenium gelatinous; paraphyses unbranched; asci oblong, unitunicate, I−, 8 spored; ascospores transversely 3-11 septate, brown, walls unequally thickened to produce lenticular locules. *Pycnidia* immersed; fulcrum exobasidial; pycnospores bacilliform, curved.
CHEMISTRY: No data.
PHYCOBIONT: Trentepohlia.
DISTRIBUTION: Queensland, New South Wales and Victoria, on bark. 7 species reported (70).

SARCOGRAPHINA MULL. ARG.: *Flora, Jena* 70, 425 (1887a)

MORPHOLOGY: Thallus crustose, uniform; ecorticate or with a poorly developed cortex of periclinal hyphae; attached by medullary or prothallial hyphae. *Ascocarp* an apothecium, elongate or irregular; immersed in raised stroma; proper exciple well developed, often with grooved labia; hypothecium pale or dark; hymenium gelatinous; paraphyses unbranched; asci oblong, unitunicate, I−, 4-8 spored; ascospores oblong to fusiform; muriform, brown. *Pycnidia* not known.
CHEMISTRY: No data.
PHYCOBIONT: Trentepohlia.
DISTRIBUTION: On bark, in Queensland. 1 species reported (8).

SARCOGYNE FLOT.: *Bot. Ztg.,* 9, 753 & 759 (1851)

MORPHOLOGY: Thallus crustose, uniform, ecorticate, poorly developed; attached by medullary or prothallial hyphae. *Ascocarp* an apothecium, adnate to sessile or sub-stipitate, disk circular; margin thin, brittle, black, proper; hypothecium dark; paraphyses rarely branched, free; asci clavate, unitunicate, thick walled, I+ pale blue, tholus I+ blue, many spored; ascospores ellipsoid to globose, simple, hyaline, thin walled, minute. *Pycnidia* immersed in the thallus or in small warts; fulcrum exobasidial; pycnospores short-cylindrical.
CHEMISTRY: (2 species) no secondary lichen substances found.
PHYCOBIONT: Myrmecia (+?Trebouxia).
DISTRIBUTION: Victoria and Western Australia. 2 species reported (30).

SCHISMATOMMA MASS.: *Ric. Lich. Crost.* 55 (1852)

MORPHOLOGY: Thallus crustose, uniform; ecorticate; attached by medullary or prothallial hyphae. *Ascocarp* a round or irregular pseudothecium; with or without a poorly developed proper exciple; hypothecium black; paraphyses reticulate; ascus clavate, bitunicate, I−, 8 spored; ascospores fusiform or elongate-oblong, transversely 4-14 septate, hyaline, sometimes curved, locules sometimes of uneven size (one much larger), thin walled. *Pycnidia* immersed; fulcrum exobasidial; pycnospores bacilliform, sometimes curved.
CHEMISTRY: (8 species) lecanoric acid (3), atranorin (1), conpsoromic acid (1), erythrin (1), lepraric acid (1), norstictic acid (1), phenolics (1), psoromic acid (1), rocellin (1).
PHYCOBIONT: Trentepohlia.
DISTRIBUTION: Queensland, New South Wales, Victoria and Tasmania, on bark and possibly leaves. 7 species reported (85).

SIPHULA FR.: *Syst. orb.* 238 (1825)

MORPHOLOGY: Thallus fruticose, upright or prostrate, attached to the substrate by rhizines, often forming turfy clumps or becoming almost coralloid, little branched, flat or terete, solid, corticate; cortex paraplectenchymatous. *Ascocarp* not known. *Pycnidia* not known.
CHEMISTRY: (21 species) baeomycesic acid (6), squamatic acid (6), hypothamnolic acid (5), decarboxythamnolic acid (3), atranorin (2), chloroatranorin (2), porphyrilic acid (2), barbatic acid (1), norstictic acid (1), siphulin (1), siphulitol (1), usnic acid (1), zeorin (1).
PHYCOBIONT: Trebouxia.
DISTRIBUTION: On dry soils and alpine peats, in Queensland, New South Wales, Victoria, Tasmania, South Australia and Western Australia. 3 species reported (25).

SPHAEROPHORUS PERS.: *Ann. Bot. (Usteri)* 7, 23 (1794)

MORPHOLOGY: Thallus fruticose, caespitose, cylindrical or flattened, solid, with a well developed cortex; cortex paraplectenchymatous. *Ascocarp* an apothecium, terminal in capitate swellings, initially closed, later opening with a thalloid exciple irregularly torn from the centre exposing the almost globose hymenium; hypothecium hyaline; paraphyses fragile, thread-like; asci clavate, unitunicate, thin walled, 8 spored; asci and paraphyses disintegrate leaving the spores in a mazaedium; ascospores globose, simple, with a dark epispore. *Pycnidia* terminal; fulcrum endobasidial; pycnospores bacilliform.
CHEMISTRY: (9 species) sphaerophorin (7), constictic acid (2), fragilin (2), isousnic acid (2), squamatic acid (2), stictic acid (2), thamnolic acid (2), hypothamnolic acid (1), protocetraric acid (1)
PHYCOBIONT: Chlorophyceae.
DISTRIBUTION: Queensland, New South Wales, Victoria and Tasmania, on bark or soil in humid regions. 3 species reported (9).

SPORASTATIA MASS.: *Geneac. Lich.* 9 (1845b)

MORPHOLOGY: Thallus crustose, uniform, ecorticate, poorly developed; attached by medullary or prothallial hyphae. *Ascocarp* an apothecium, immersed; disk circular; margin poorly developed to absent, if present thin and black, proper; hypothecium dark; paraphyses rarely branched, free; asci clavate, unitunicate, thick walled, I+ pale blue, tholus I+ blue, many spored; ascospores ellipsoid to globose, simple, hyaline, thin walled, minute. *Pycnidia* immersed in the thallus or in small warts; fulcrum exobasidial; pycnospores short-cylindrical.
CHEMISTRY: No data.
PHYCOBIONT: Myrmecia (? + Trebouxia).
DISTRIBUTION: On alpine rocks, New South Wales. 1 species reported (4).

SPOROPODIUM MONT. em. R. SANT.: *Symb. bot. Upsal.* 12, 509 (1952)

MORPHOLOGY: Thallus crustose, uniform, ecorticate; attached by medullary or prothallial hyphae; prothallus often well developed especially at the margins. *Ascocarp* an apothecium, initially immersed, ultimately sessile; well developed proper exciple; hypothecium pale or dark; paraphyses branched and somewhat anastomosing; ascus clavate, unitunicate, thick walled, 1 spored; ascospore muriform, hyaline, thin walled. *Pycnidia* unknown.
CHEMISTRY: (4 species) pulvinic acid (2), pulvinic dilactone (2), vulpinic acid (2), calycin (1), 3-0-methyl-2,5,7-trichloronorlichexanthone (1), 2,5,7-trichloronorlichexanthone (1).
PHYCOBIONT: ?Trebouxia.
DISTRIBUTION: Queensland and New South Wales, on leaves. 2 species reported (9).

SPILONEMA BORNET: *Mem. Soc. Sci. Nat. Cherbourg* 4, 226 (1856)

MORPHOLOGY: Thallus dwarf fruticose, of hyphae extending longitudinally within the algal sheath, undifferentiated. *Ascocarp* an apothecium, terminal on branches; disk flat to convex; exciple thin proper, soon disappearing; hypothecium brown; paraphyses unbranched, thick; asci oblong-clavate, unitunicate, I+, I+ blue tholus, 8 spores; ascospores oblong-ellipsoid, simple; hyaline. *Pycnidia* immersed; fulcrum endobasidial; pycnospores bacilliform.
CHEMISTRY: No data.
PHYCOBIONT: Stigonema.
DISTRIBUTION: On maritime rocks, Western Australia. 1 species reported (4).

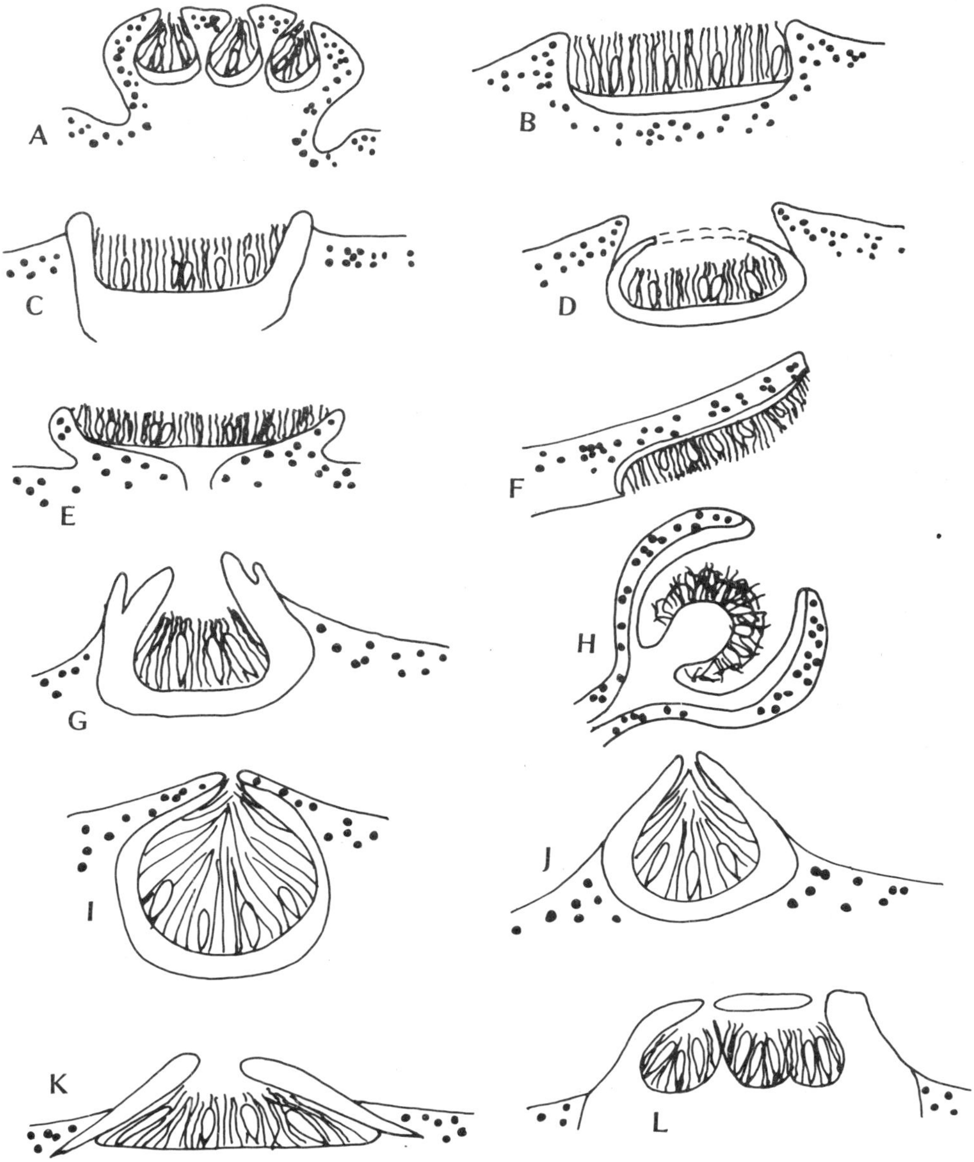

FIGURE 4 Variation in ascocarp structure and placement when viewed in section. (A) apothecia of *Pertusaria* immersed in thalloid warts; (B) immersed apothecium of *Aspicilia* showing the algal layer continuing below the hymenium, making this a thalloid margin (technically crypto-lecanorine); (C) apothecium of *Rhizocarpon,* which in contrast with *Aspicilia* does not have algae below the hymenium and therefore has a proper margin; (D) apothecium of *Thelotrema* showing the distinct proper and thalloid margins, with the originally closed proper covering dotted in; (E) sessile thalloid apothecium of *Caloplaca*; (F) apothecium of *Nephroma* on the lower surface of the lobe, apparently emarginate; (G) partly immersed apothecium of *Graphis* with grooved labia or margins; (H) apothecium of *Sphaerophorus* immersed in the tip of a lobe showing the central columella which supports the asci, etc.; (I) immersed perithecium of *Dermatocarpon*; (J) partly immersed perithecium of *Pyrenula*; (K) partly immersed dimidiate or conical perithecium-like pseudothecium of *Arthopyrenia,* in which no wall is formed below the ascocarp; (L) compound perithecium-like pseudothecium of *Mycoporellum,* showing several hymenia in a partly divided chamber. Paraphyses and paraphysis-like structures are represented by lines forming the hymenium, algae by black dots.

SQUAMARINA POELT: *Mitt. bot. Stamml. Munch.* 5, 19-20 (1958)

MORPHOLOGY: Thallus squamulose, attached to the substrate by medullary hyphae, cortex developed on both surfaces. *Ascocarp* an apothecium, round, sessile; thalloid margin thin; paraphyses unbranched; hypothecium pale; asci clavate, unitunicate, tholus I+ blue, 8 spored; ascospores oblong-ellipsoid, simple, hyaline, thin walled. *Pycnidia* immersed.
CHEMISTRY: (9 species) psoromic acid (7), usnic acid (5), parellic acid (2).
PHYCOBIONT: Trebouxia.
DISTRIBUTION: Queensland, New South Wales, Victoria and Tasmania. ?5 species reported (18).

STENOCYBE NYL.: *Bot. Not.* 84 (1854)

MORPHOLOGY: Thallus thin, crustose, poorly developed or absent. *Ascocarp* an apothecium terminal on a usually well developed stipe; at first closed by the exciple, but eventually opening; usually dark brown or black; paraphyses unbranched, slender; asci cylindrical, thick walled, unitunicate, 8 spored; ascospores ellipsoid to oblong-ellipsoid, 3-7 transversely septate, brown. *Pycnidia* not known.
CHEMISTRY: No data.
PHYCOBIONT: ?Trebouxia-like or absent.
DISTRIBUTION: Tasmania. ?1 species reported (5).

STEREOCAULON SCHREB.: *Gener. Plant.* 768 (1791)

MORPHOLOGY: Thallus more or less crustose to squamulose, producing thalline pseudopodetia, the rudimentary horizontal thallus usually disappearing early; pseudopodetia solid, somewhat branched, fruticose in appearance, ecorticate, bearing cylindrical lateral ecorticate structures (phyllocladia) containing algae; cephalodia on pseudopodetia are irregularly globose. *Ascocarp* an apothecium, terminal on pseudopodetia; disk flat to convex, red-brown to black; exciple proper, soon disappearing; hypothecium hyaline; paraphyses unbranched; ascus clavate, unitunicate, I+ blue, tholus I+ blue, 8 spored; ascospores fusiform to acicular, transversely 3-7 septate, hyaline, thin walled. *Pycnidia* terminal or lateral, immersed; fulcrum exobasidial; pycnospores thread-like to cylindrical, straight or curved.
CHEMISTRY: (74 species) atranorin (56), lobaric acid (33), stictic acid (22), norstictic acid (16), porphyrilic acid (9), perlatolic acid (5), anziaic acid (4), α-methyl-ether-salazinic acid (3), miriquidic acid (3), salazinic acid (3), psoromic acid (2), tetrahydroxy fatty acids (2), thamnolic acid (2), colensoinic acid (1), constictic acid (1), divaricatic acid (1), fumarprotocetraric acid (1), lecanoric acid (1), parietin (1), β-setosterol (1), squamatic acid (1), ventosic acid (1).
PHYCOBIONT: Trebouxia with *Nostoc* or *Stigonema* cephalodia.
DISTRIBUTION: Queensland, New South Wales, Victoria and Tasmania, on soil in humid areas. 10 species reported (120).

STICTA SCHREB.: *Gener. Plant.* 768 (1791)

MORPHOLOGY: Thallus foliose, large, loosely adnate or ascending; both surfaces paraplectenchymatously corticate with anticlinal hyphae; internal cephalodia sometimes present below the upper cortex; attached to the substrate by short or long rhizoids; rhizoids sometimes almost forming a tomentum which may be more or less confined to a central

umbilicus-like area; lower cortex pierced by cyphellae. *Ascocarp* an apothecium, marginal or laminal, sessile to sub-stipitate, usually with a thalloid exciple; disk round, flat to concave; hypothecium pale or coloured; paraphyses unbranched; asci clavate to ovate-clavate, unitunicate, I+ blue, 8 spored; ascospores oblong-fusiform to acicular, transversely 1-7 septate, hyaline or brown, thin walled. *Pycnidia* warty, immersed, marginal or laminal; fulcrum endobasidial; pycnospores bacilliform or constricted in the middle.
CHEMISTRY: (13 species) pulvinic dilactone (2), stictalbin (2), stictinin (2), triterpenes (2), 7β-acetoxy-22-hydroxyhopane (1), anthraquinones (1), calycin (1), 15α,22-dihydroxyhopane (1), gyrophoric acid (1), lecanoric acid (1), norstictic acid (1), polyporic acid (1), stictic acid (1).
PHYCOBIONT: *Nostoc* or *Palmella* with *Nostoc* cephalodia.
DISTRIBUTION: Queensland, New South Wales Victoria and Tasmania, on rocks or bark in humid regions. 50 species reported (200).

STRIGULA FR.: *K. svenska Vetensk. Akad. Handl.* 1821, 323 (1821)

MORPHOLOGY: *Thallus* crustose, ecorticate, with minutely lobed margins; sub-cuticular. *Ascocarp* a perithecium, either naked or immersed in the thallus; erect; without hymenial algae; paraphyses unbranched; asci cylindrical or oblong-clavate, bitunicate, thin walled, I−, 8 spored; ascospores ellipsoid to fusiform, transversely 1-3 septate, hyaline, thin walled. *Pycnidia* protruding, convex; fulcrum exobasidial; pycnospores ellipsoid to fusiform; stylospores 1-9 septate, bacilliform or filiform.
CHEMISTRY: No data.
PHYCOBIONT: *Cephaleuros.*
DISTRIBUTION: Queensland and New South Wales, on leaves in rainforests. 7 species reported (12).

SYNALISSA FR.: *Syst. orb.* 297 (1825)

MORPHOLOGY: *Thallus* minutely fruticose, lobes erect, little branched, ecorticate, undifferentiated; composed of little branched hyphae in the algal gel; attached to the substrate by rhizines. *Ascocarp* an apothecium, terminal, immersed in the lobe tips; initially closed, later open with a thalloid margin; hymenium gelatinous; hypothecium hyaline to brown; paraphyses unbranched; asci clavate, unitunicate, thin walled, I+ blue, 8-32 spored; ascospores ellipsoid to spherical, simple, hyaline, thin walled. *Pycnidia* immersed; fulcrum endobasidial; pycnospores bacilliform.
CHEMISTRY: No data.
PHYCOBIONT: *Gloeocapsa.*
DISTRIBUTION: Victoria, Tasmania, South Australia and Western Australia, on calcareous soils. ?2 species reported (5).

TAPELLARIA MULL. ARG. em R. SANT.: *Symb. bot. Upsal.* 12, 494 (1952)

MORPHOLOGY: *Thallus* crustose, uniform, ecorticate; attached to the substrate by medullary or prothallial hyphae. *Ascocarp* an apothecium, sessile or immersed; disk round with a well developed black proper exciple; hypothecium purple-brown; without epithecial algae; paraphyses richly reticulate, enveloping the asci; asci clavate, unitunicate thick walled, 1-8 spored; ascospores oblong, transversely septate or muriform, hyaline, large. *Pycnidia* rare, sessile or partly immersed, globose; fulcrum endobasidial; pycnospores bacilliform.

CHEMISTRY: (1 species) arthothelin (1), thuringione (1).
PHYCOBIONT: ?*Trebouxia.*
DISTRIBUTION: Queensland, on leaves in humid forests. 1 species reported (8).

TELOSCHISTES NORM.: *Nytt. Mag. Naturv.* 7, 228 (1853)

MORPHOLOGY: *Thallus* fruticose or almost foliose, corticate on all surfaces; cortex of longitudinal agglutinate hyphae; usually without rhizines but often ciliate; attached to the substrate by rhizoidal hyphae. *Ascocarp* an apothecium, marginal or laminal; sessile; disk round, concave to flat, yellow or orange; margin thalloid; epithecium with algae; hypothecium pale; paraphyses unbranched; ascus clavate, unitunicate, I+ blue, tholus I+ blue, 8 spored; ascospores oblong-ellipsoid, polaribilocular, or sometimes 3 septate, thick walled, hyaline. *Pycnidia* globose; fulcrum endobasidial; pyncnospores bacilliform, straight.
CHEMISTRY: (11 species) parietin (11), teloschistin (7), emodin (6), fallacinal (6), parietinic acid (6), xanthorin (6), erythroglaucin (5), salazinic acid (1).
PHYCOBIONT: *Trebouxia.*
DISTRIBUTION: On bark and rocks, especially in the sub-humid and humid areas of the southern half of the region. 7 species reported (30).

THAMNOLIA ACH. ex SCHAER.: *Enum. lich. eur.* 243 (1850)

MORPHOLOGY: *Thallus* fruticose, upright or prostrate, more or less cylindrical, little branched, inflated, tapering to the ends; corticate on all surfaces; cortex paraplectenchymatous of anticlinal hyphae; unattached. *Ascocarp* not known. *Pycnidia* not known.
CHEMISTRY: (3 species) atranorin (1), baeomycesic acid (1), decarboxythamnolic acid (1), squamatic acid (1), thamnolic acid (1).
PHYCOBIONT: *Trebouxia.*
DISTRIBUTION: On soils, in sub-alpine and alpine areas of Victoria, New South Wales and Tasmania. 3 species reported (3).

THELIDIUM MASS.: *Framment. lich.* 15 (1855)

MORPHOLOGY: *Thallus* crustose, uniform, ecorticate. *Ascocarp* a perithecium, sessile on the thallus; without algae in the hymenium; paraphyses gelatinizing; asci clavate or inflated, unitunicate, thick walled, I−, 8 spored; ascospores ellipsoidal to ovoid, transversely 1-3 septate, hyaline. *Pycnidia* not known.
CHEMISTRY: No data.
PHYCOBIONT: *Pleurococcus* and others?
DISTRIBUTION: Tasmania, on soil. ?1 species reported (100).

THELOTREMA ACH.: *Meth. Lich.* 130 (1803)

MORPHOLOGY: *Thallus* crustose, epi- or endophloic, uniform; ecorticate or with a woolly poorly developed cortex; attached by medullary or prothallial hyphae. *Ascocarp* an apothecium, immersed to sessile in warts; disk round or elongate, punctiform or narrow; well developed proper exciple initially closed over the disk, later tearing and opening then disintegrating, leaving the disk free but surrounded by a thalloid margin; paraphyses unbranched; asci cylindrical-clavate, unitunicate, thick walled, I−, 8 spored; ascospores ellipsoid or oblong-ellipsoid, hyaline, muriform, wall thickness unequal resulting in spherical or lenticular locules.

Pycnidia, fulcrum exobasidial.
CHEMISTRY: (34 species) 4-0-methyldenotatic acid (6), psoromic acid (6), constictic acid (5), fumarprotocetraric acid (5), hypoprotocetraric acid (5), stictic acid (5), conpsoromic acid (4), 4-0-methylhypoprotocetraric acid (2), norstictic acid (2), protocetraric acid (2), salazinic acid (1).
PHYCOBIONT: Trentepohlia.
DISTRIBUTION: Queensland, New South Wales, Victoria, Tasmania and Western Australia, on bark. 20 species reported (200).

THYREA MASS.: *Flora, Jena* 39, 210 (1856a)

MORPHOLOGY: Thallus foliose, monophyllous to entire, or sinuose and incised, attached at a central point; gelatinous; composed of hyphae ramifying through the algal gelatin; ecorticate. *Ascocarp* an apothecium, immersed to adnate; the disk deeply concave to flat, with a thick thalloid margin which may more or less close the disk; hypothecium hyaline to brown; paraphyses unbranched; ascus clavate, unitunicate, thin walled, ?I+, 8-24 spored; ascospores spherical to ellipsoid, simple, hyaline, thin walled. *Pycnidia* immersed; fulcrum exobasidial; pycnospores oblong.
CHEMISTRY: No data.
PHYCOBIONT: Gloeocapsa.
DISTRIBUTION: Western Australia, on (? calcareous) rocks or soil. ?1 species reported (20).

THYSANOTHECIUM MONT. et BERK.: *London J. Bot.* 5, 257 (1846)

MORPHOLOGY: Thallus granular to sub-squamulose; upper surface corticate; attached to the substrate by rhizoidal hyphae. *Ascocarp* an apothecium borne on podetia; podetia erect, usually unbranched, with a cortex, the medulla solid with longitudinal cords extending behind the hymenium, the algae in a sub-cortical layer; apothecia at first terminal with a raised proper margin, eventually lateral on bifacially flattened area without a developed margin and irregular in outline, often somewhat lobed; hypothecium pale; paraphyses little branched; ascus clavate, unitunicate, I+ blue, tholus I+ blue, 8 spored; ascospores ellipsoid, simple or 1 septate, hyaline, thin walled. *Pycnidia* not known.
CHEMISTRY: (3 species) divaricatic acid (3), usnic acid (2).
PHYCOBIONT: Trebouxia.
DISTRIBUTION: On charcoal and soil, throughout humid parts of southern Australia. 2 species reported (3).

TOMASELLIA MASS.: *Flora, Jena* 39, 283 (1856)

MORPHOLOGY: Thallus crustose, uniform, epi- or endophloic, ecorticate. *Ascocarp* a perithecium-like ascolocular pseudothecium, immersed in groups in stroma; ostiole apical, horizontal; paraphyses reticulate; ascus bitunicate, thick walled, I−, 8 spored; ascospores ovoid to acicular, transversely 2-16 septate, hyaline, thin walled. *Pycnidia* not known.
CHEMISTRY: No data.
PHYCOBIONT: Trentepohlia.
DISTRIBUTION: Queensland, on bark in humid forests. 4 species reported (23).

TONINIA MASS.: *Ric. Lich. Crost.* 107 (1852)

MORPHOLOGY: Thallus crustose to small squamulose or almost foliose or fruticose, sometimes inflated or lobed at the margin; cortex of periclinal hyphae; attached to the substrate by medullary hyphae. *Ascocarp* an apothecium, sessile; disk round; margin proper, dark, often disappearing; hypothecium pale or dark; paraphyses unbranched; ascus clavate, unitunicate, thin walled, I+ blue, tholus I+ blue, 8 spored; ascospores oblong to acicular, transversely 1-7 septate, hyaline, thin walled. *Pycnidia* immersed; fulcrum exobasidial; pycnospores filiform or acicular, curved.
CHEMISTRY: (3 species) atranorin (2), norstictic acid (1), stictic acid (1).
PHYCOBIONT: Trebouxia.
DISTRIBUTION: Throughout the region, especially on soils from arid to alpine zones. 7 species reported (80).

TRAPELIA CHOISY: *Bull. mens. Soc. linn. Lyon.* 18, 112 (1949)

MORPHOLOGY: Thallus crustose to minutely squamulose, sometimes with a thin cortex, attached to the substrate by medullary hyphae. *Ascocarp* an apothecium, sessile to adnate, round; usually with a brown proper margin; hypothecium hyaline; paraphyses thin, often branched and reticulate; ascus clavate to cylindrical, unitunicate, thin walled, I−, 8 spored; ascospores ellipsoid, simple, hyaline. *Pycnidia* immersed; fulcrum exobasidial; pycnospores oblong to filiform.
CHEMISTRY: (2 species) gyrophoric acid (2), lecanoric acid (2).
PHYCOBIONT: Coccoid chlorophyceae.
DISTRIBUTION: New South Wales, Victoria, South Australia and Western Australia, on compacted earth. 1 species reported (12).

TREMOTYLIUM NYL.: *Bull. Soc. linn. Normandie Ser. 2,* 2, 513 (1868)

MORPHOLOGY: Thallus crustose, uniform, ecorticate; attached to the substrate by medullary hyphae. *Ascocarp* an apothecium, immersed in stroma; disk irregular, narrow, concave; well developed proper exciple with a thalloid margin; paraphyses branched, reticulate; ascus clavate, unitunicate, 8 spored; ascospores ovoid to oblong, hyaline or brown, muriform, walls unevenly thickened resulting in globose locules. *Pycnidia* not known.
CHEMISTRY: No data.
PHYCOBIONT: Trentepohlia.
DISTRIBUTION: Queensland, on bark. 2 species reported (6).

TRICHARIA FEE em. R. SANT.: *Symb. bot. Upsal.* 12(1), 376 (1952)

MORPHOLOGY: Thallus crustose, uniform; ecorticate or with a cartilaginous cortex; usually with black or white erect hairs. *Ascocarp* an apothecium, sessile; disk round with a proper margin; hypothecium hyaline; paraphyses reticulate; ascus clavate, unitunicate, thick walled, 1-8 spored; ascospores ovoid, transversely septate or muriform, hyaline. *Pycnidia* not known.
CHEMISTRY: No data.
PHYCOBIONT: Coccoid chlorophyceae.
DISTRIBUTION: Queensland, on leaves in humid forests. 1 species reported (10).

TRICHOTHELIUM MULL. ARG. em R. SANT.: *Symb. bot. Upsal.* 12(1), 266 (1952)

MORPHOLOGY: Thallus crustose, uniform, ecorticate. *Ascocarp* a perithecium, sessile; the ostiole surrounded by stiff horizontal bristles; paraphyses unbranched; ascus clavate, thin walled, the endoascus reduced to a thickening at the tip, I−, 8 spored; ascospores acicular, transversely many septate or muriform, thin walled. *Pycnidia* not known.
CHEMISTRY: No data.
PHYCOBIONT: Phycopeltis.
DISTRIBUTION: Queensland, on leaves in humid forests. 1 species reported (7).

TRIMMATOTHELE NORM. ex BLOMBERG et FORSSELL: *Enum. Pl. Scand.* 100 (1880)

MORPHOLOGY: Thallus crustose, epi- or endophloic, uniform, ecorticate. *Ascocarp* a perithecium, sessile or partly immersed; ostiole erect, minute; paraphyses absent; without hymenial algae; asci obscure, unitunicate, thin walled, many spored; ascospores minute, ovoid or ellipsoid, simple, hyaline. *Pycnidia* not known.
CHEMISTRY: No data.
PHYCOBIONT: Coccoid chlorophyceae.
DISTRIBUTION: Queensland, on trunks of palms. 1 species reported (5).

TRYPETHELIUM SPRENG.: *Anleit. Kenn. Gewachse* 309 (1805)

MORPHOLOGY: Thallus crustose, uniform, epi-or endophloic; ecorticate or weakly corticate with periclinal hyphae. *Ascocarp* a perithecium; immersed in groups in stroma; paraphyses reticulate; asci cylindrical-clavate, bitunicate, I−, 8 spored; ascospores cylindrical to fusiform, transversely 3-17 septate, hyaline, walls unequally thickened resulting in lenticular or globose locules. *Pycnidia* immersed in stroma; fulcrum exobasidial; pycnospores cylindrical or constricted in the middle.
CHEMISTRY: (5 species) parietin (4), emodin (1), lichexanthone (2), norlichexanthone (1), secaleconic acid (1).
PHYCOBIONT: Trentepohlia.
DISTRIBUTION: Queensland, on bark. 12 species reported (100).

TYLOPHORON NYL.: *Bot. Ztg.* 20, 279 (1862)

MORPHOLOGY: Thallus crustose or absent, effuse, ecorticate. *Ascocarp* an apothecium, initially immersed in thalline warts, later sessile, cylindrical or cup-shaped; disk open; proper and thalloid exciples present; hypothecium hyaline to dark, produced concially downwards; asci cylindrical, thin walled, unitunicate, disintegrating at maturity to release the spores in a mazaedium of ascal and hyphal fragments, 8 spored; ascospores globose to fusiform, 1 or 2 septate, hyaline to brownish, with a thickened wall. *Pycnidia* immersed; fulcrum exobasidial; pycnospores acicular, straight.
CHEMISTRY: No data.
PHYCOBIONT: Trentepohlia.
DISTRIBUTION: Queensland, on bark. 1 species reported (12).

UMBILICARIA HOFFM.: *Descr. adumbr. Lich.* 1, 9 (1789)

MORPHOLOGY: Thallus foliose, both surfaces corticate; cortex paraplectenchymatous attached by an umbilicus, although non-functional rhizines may be found on the lower surface; margins often markedly lacerate. *Ascocarp* an apothecium, sessile; sometimes in depressions or almost stipitate; surrounded by a proper margin; often with a gyrose disk; hypothecium brown; paraphyses unbranched; ascus clavate, unitunicate, I+ blue, ascal gelatin I+ violet, 1-8 spored; ascospores ellipsoid to oblong ellipsoid, simple to muriform; hyaline to brown. *Pycnidia* wart-like, fulcrum endobasidial; pycnospores bacilliform.
CHEMISTRY: (19 species) gyrophoric acid (18), umbilicaric acid (5), lecanoric acid (2), norstictic acid (1), stictic acid (1).
PHYCOBIONT: Trebouxia.
DISTRIBUTION: New South Wales, Victoria and Tasmania, on sub-alpine or alpine rocks. 7 species reported (45).

USNEA (HILL.) WIGG.: *Prim. fl. Holsat.* 90 (1780)

MORPHOLOGY: Thallus fruticose, erect or pendulous, corticate; cortex horny, brittle, of agglutinate hyphae; medulla woolly; central axis cartilaginous, solid or with a hollow centre, of longitudinal agglutinated hyphae; attached to the substrate by a basal disk; thallus straw-yellow to grey-green or red. *Ascocarp* an apothecium, lateral or terminal on branches; well developed thalloid margin; disk round, pale; hypothecium pale; paraphyses branched, agglutinate; asci clavate to cylindrical, untunicate, thick walled, I+ blue, tholus I+ blue, 8 spored; ascospores ellipsoid to spherical, simple, hyaline, thin walled. *Pycnidia* immersed; fulcrum exobasidial; pycnospores fusiform to acicular.
CHEMISTRY: (135 species) usnic acid (131), salazinic acid (50), norstictic acid (37), stictic acid (20), barbatic acid (17), diffractaic acid (13), protocetraric acid (12), thamnolic acid (9), atranorin (6), constictic acid (6), psoromic acid (6), fumarprotocetraric acid (4), galbinic acid (4), caperatic acid (3), connorstictic acid (3), conpsoromic acid (2), decarboxythamnolic acid (2), barbatolic acid (1), chloroatranorin (1), divaricatic acid (1), evernic acid (1), lobaric acid (1), obtusatic acid (1), physodalic acid (1), squamatic acid (1).
PHYCOBIONT: Trebouxia.
DISTRIBUTION: On wood, bark and rock, in all states. 40 species reported (600).

VERRUCARIA WIGG.: *Prim. fl. Holsat.* 85 (1780)

MORPHOLOGY: Thallus crustose, often powdery; sometimes endolithic; ecorticate; prothallus may be prominent. *Ascocarp* a perithecium; immersed to sessile; ostiole punctiform or radially cracked, paraphyses disappearing early; without hymenial algae; asci ovoid to ovoid-clavate, unitunicate, thick walled, I−, 8 spored; ascospores oblong-ellipsoid to globose, simple, hyaline to brown, thin walled. *Pycnidia* immersed; fulcrum endobasidial; pycnospores acicular, straight or curved.
CHEMISTRY: (1 species) no characteristic lichen substances found.
PHYCOBIONT: Myrmecia, Coccobotrys, Pleurococcus, Heterococcus, Pseudopleurococcus.
DISTRIBUTION: Victoria, South Australia and Western Australia, on rock, 6 species reported (300).

XANTHOPARMELIA (VAIN.) HALE: *Phytologia* 28, 485 (1974b)

MORPHOLOGY: *Thallus* foliose, lobed, radiate, appressed to somewhat ascending; both surfaces corticate; lobes narrow, eciliate; attached to the substrate by simple rhizines; upper surface yellow-green. *Ascocarp* an apothecium, sessile, laminal; disk concave, brown, circular; margin thalloid; hypothecium pale; paraphyses branched; ascus clavate, unitunicate, I+ blue, tholus I+ blue, 8 spored; ascospores ellipsoid, simple, hyaline, thin walled. *Pycnidia* immersed; fulcrum endobasidial, bayonet-like; pycnospores bacilliform.
CHEMISTRY: (51 species) usnic acid (51), salazinic acid (12), hypoprotocetraric acid (7), loxodin (7), norlobaridone (7), 4-0-methyldenotatic acid (6), norstictic acid (6), fumarprotocetraric acid (4), succinprotocetraric acid (4), connorstictic acid (3), lecanoric acid (3), stictic acid (3), alectoronic acid (2), atranorin (2), conpsoromic acid (2), notatic acid (2), protocetraric acid (2), psoromic acid (2), barbatic acid (1), constictic acid (1), diffractaic acid (1), 4-0-methylhypoprotocetraric acid (1), olivetoric acid (1), physodalic acid (1), "quintaria" unknown (1), rhizonic acid (1).
PHYCOBIONT: *Trebouxia.*
DISTRIBUTION: Throughout the region, on rocks and soil. ?30 species reported (?100).

XANTHORIA (FR.) TH. FR.: *Nova Acta R. Soc. Scient. upsal. Ser 3,* 3, 166 (1861)

MORPHOLOGY: *Thallus* foliose, more or less appressed, lobed, radiate, golden-yellow; both surfaces corticate, paraplectenchymatous of anticlinal hyphae; attached to the substrate by rhizines. *Ascocarp* an apothecium, laminal, sessile; margin thalloid; disk round; hypothecium pale, paraphyses unbranched or little branched; ascus clavate, unitunicate, I+ blue, tholus I+ blue, 8 spored; ascospores ellipsoid, polari-bilocular, hyaline. *Pycnidia* immersed in thalloid warts; fulcrum endobasidial; pycnospores bacilliform.
CHEMISTRY: (14 species) parietin (14), fallacinal (11), teloschistin (11), emodin (10), parietinic acid (9), erythroglaucin (5), xanthorin (4), citeorosein (2), emodic acid (2), emodinal (1), erythroglaucinic acid (1).
PHYCOBIONT: *Trebouxia.*
DISTRIBUTION: Throughout the southern half of the region, on bark and rock, 2 species reported (15).

XYLOGRAPHA FR.: *Summ. veg. Scand.* 372 (1846).

MORPHOLOGY: *Thallus* crustose, little differentiated, attached to the substrate by medullary hyphae. *Ascocarp* an apothecium, sessile to immersed, minute (c.o. 1mm diam.), disk round to irregular; margin absent or little developed; hypothecium pale, paraphyses simple; asci with an apical wall thickening, I + blue then green, unitunicate, without an I + blue tholus, eight spored; ascospores simple, hyaline, thin walled. *Pycnidia* not known.
CHEMISTRY: (3 species) norstictic acid (2), atranorin (1), stictic acid (1).
PHYCOBIONT: Protococcoid.
DISTRIBUTION: On dead wood in Victoria. 1 species reported (5).

Aids for Species Determination

Few genera have been revised for the Australian region, and some worldwide revisions have been made using very little Australian material. To name a lichen to species therefore presents severe problems for many genera in most or all of Australia. The only regional lichen flora which is of use over a wide area is that of South Australia prepared by Filson and Rogers (1979), and that is incomplete. To assist with species identifications a list of handbooks and publications is provided below. The list has three sections: accounts of the lichen flora in some restricted areas of Australia, lichen floras from other regions which are likely to be useful and a list of other publications to help nomenclature within the named genera. Many lichens are cosmopolitan, or at least widespread within a climatic zone. Reference to a flora or revision from a similar climatic zone is therefore likely to assist in the naming of Australian collections. If the Australian material is of a different species from those covered in any given account of a genus, at least an idea of the characters useful in separating the species can be gleaned from the literature, as can an idea of the affinity of the material in hand.

1. REGIONAL ACCOUNTS OF THE AUSTRALIAN LICHENS

Filson R.B. and Rogers, R.W. (1979) *The Lichens of South Australia*, (Government Printer, Adelaide). A flora with keys to families, genera and species, descriptions, glossary, etc. Crustose genera are not usually treated at species level. Useful in western Victoria, western and central New South Wales and Western Australia as well as South Australia.

Rogers, R.W. (1974) "Lichens from the T.G.B. Osborn Vegetation Reserve at Koonamore in arid South Australia", *Trans. R. Soc. S. Aust.* 98, 113-23. Keys and descriptions of all species found in this small arid reserve. Useful throughout arid southern Australia.

Rogers, R.W. and Lange, R.T. (1972) "Soil surface lichens in arid and sub-arid south-eastern Australia. I. Introduction and floristics", *Aust. J. Bot.* 20, 197-213. Key to soil surface species, useful throughout arid southern Australia.

Stevens, G.N. and Rogers, R.W. (1979) "The macrolichen flora from the mangroves of Moreton Bay", *Proc. R. Soc. Qd.* 90, 33-49. Keys and descriptions of species with some plates, useful in mangroves from Sydney to Bundaberg, but of diminishing use beyond.

2. USEFUL FLORAS FROM OTHER REGIONS

Duncan, U.K. (1970) *Introduction to British Lichens* (Buncle, Arbroath). Useful, especially for crustose and alpine-sub-alpine genera.

Fink, B (1935) *The Lichen Flora of the United States* (Michigan U.P., Ann Arbor). A very dated account but still helpful especially in crustose genera.

Galun, M. (1970) *The Lichens of Israel* (Israel Academy of Sciences and Humanities, Jerusaleum). Keys and descriptions of species, well illustrated, good for Mediterranean climates.

Hale, M.E. (1969) *How to Know the Lichens* (Brown, Dubuque). Very helpful with foliose and fruticose genera; every species illustrated.

Martin, W. and Child, J. (1972) *Lichens of New Zealand* (Reed and Reed, Wellington). An incomplete illustrated account with keys to common species, of value especially in Victoria and Tasmania.

Poelt, J. (1969) *Bestimmungsschlussel Europaischer Flechten* (J. Cremer, Vaduz). Keys to genera and species, useful especially in cooler areas.

Vainio, E.A. (1890) "Etude sur la classification naturelle et la morphologie des Lichens du Bresil", *Acta Soc. Fauna Flora fenn.* 7(1) i-247, 7(2) 1-256. An account of tropical lichens still unsurpassed in breadth and detail. The text is in Latin, and is without keys, but the clarity of descriptions make the work usable in tropical and sub-tropical areas. It is likely to be shelved in libraries as a book *Lichens du Bresil* with the author listed as Wainio, an alternative spelling he sometimes used.

3. ACCOUNTS OF GENERA

Acarospora:	Magnusson (1929); Weber (1968).
Alectoria:	Brodo and Hawksworth (1977); Jorgensen (1972); Motyka (1964).

Anaptychia:	Kurokawa (1962, 1973); Swinscow and Krog (1976a).
Athracothecium:	Keissler (1937a); Muller-Argoviensis (1895).
Anzia:	Galloway (1978) Bratt, Blackman and Cashin (1976).
Arctomia:	Henssen (1969).
Arthonia:	Redinger (1936a); Santesson (1952).
Arthopyrenia:	Keissler (1937a).
Arthothelium:	Redinger (1936a).
Aspicilia:	Weber (1967a).
Aspidothelium:	Santesson (1952).
Aulaxina:	Santesson (1952).
Bacidia:	Malme (1935); Santesson (1952), Zahlbruckner (1941).
Baeomyces:	Thomson (1967).
Biatorella:	James (1973); Magnusson (1935c).
Bombyliospora:	Malme (1923).
Bottaria:	Malme (1924); Muller- Argoviensis (1895).
Buellia:	Imshaug (1955); Magnusson (1953); Malme (1928); Sheard (1964).
Bulbothrix::	Hale (1976b).
Byssoloma:	Santesson (1952).
Calicium:	Keissler (1937d); Tibell (1975).
Caloplaca:	Alon and Galun (1971); Malme (1926); Wade (1965).
Campylothecium:	Malme (1924).
Candelaria:	Almborn (1966); Poelt (1974).
Candelariella:	Hakulinen (1954).
Catillaria:	Malme (1923); Santesson (1952).
Catinaria:	Vainio (1922).
Cetraria:	Esslinger (1973); Karnefelt (1977).
Cetrelia:	Culberson and Culberson (1968).
Chaenotheca:	Keissler (1937d); Tibell (1975).
Chondropsis:	Bibby (1955)
Cladia:	Filson (1970); Martin (1965).
Cladina:	Ahti (1961); Thomson (1968).
Cladonia:	Martin (1958); Thomson (1968).
Clathroporina:	Keissler (1937a).
Coccocarpia:	Arvidsson and Galloway (1979); Santesson (1952); Swinscow and Krog (1976b).
Coccotrema:	Brodo (1973); Oshio (1968).
Coenogonium:	Dodge (1933); Santesson (1952a); Vainio (1921).
Collema:	Degelius (1954, 1974).
Coniocybe:	Keissler (1937d).
Conotremopsis:	Vezda (1977).
Crocynia:	Hue (1924).
Cyphelium:	Tibell (1971); Weber (1967b).
Dermatocarpon:	Magnusson (1940); Swinscow and Krog (1975c); Zschacke (1933b).
Dermatina:	Keissler (1937c); Reidl (1964).
Dictyonema:	Parmasto(1978)
Dimerella:	Lettau (1937); Santesson (1952).
Diplogramma:	Muller Argoviensis (1891).
Diploschistes:	Lettau (1937); Magnusson (1955).
Dirinaria:	Awasthi (1975).
Dirinastrum:	Muller Argoviensis (1893).

Echinoplaca:	Santesson (1952).
Endocarpon:	Magnusson (1940); Zschacke (1933b).
Enterographa:	Redinger (1936b, 1938); Santesson (1952).
Ephebe:	Henssen (1963a).
Erioderma:	Keuck (1977).
Evernia:	Bird (1974).
Fistulariella:	Bowler and Rundel (1977); Krog and Swinscow (1976b).
Fulgensia:	Poelt (1965).
Gylphis:	Redinger (1933).
Gomphyllus:	Frey (1933).
Graphina:	Hayward (1977); Nakanishi (1966); Wirth and Hale (1963, 1978).
Graphis:	See *Graphina.*
Gyalecta:	Malme (1934).
Gyalectidium:	Santesson (1952).
Gymnoderma:	Yoshimura (1973); Yoshimura and Sharp (1968).
Gymnographa:	Muller Argoviensis (1887)
Haematomma:	Culberson (1964).
Heppia:	Wetmore (1970)
Heterodea:	Filson (1978)
Heterodermia:	See *Anaptychia.*
Hypogymnia:	Elix (1980).
Hypotrachyna:	Hale (1975a); Hale and Kurokawa (1964); Krog and Swinscow (1979).
Icmadophila:	Frey (1933).
Laurera:	Malme (1924).
Lecanactis:	Lettau (1937).
Lecanora:	Imshaug and Brodo (1966); Poelt (1952, 1958).
Lecidea:	Hertel (1967, 1968, 1969b); Zahlbruckner (1941).
Lepraria:	Hue (1924).
Leprocaulon:	Lamb and Ward (1974).
Leptogium:	Dodge (1933); Sierk (1964).
Leptotrema:	Hale (1978).
Lichina:	Henssen (1963a).
Lobaria:	Jordan (1973); Yoshimura (1971).
Lopadium:	Kurokawa and Kashiwadani (1977); Santesson (1952).
Mallotium:	Awasthi and Akhtar (1977); Jorgensen (1975); Sierk (1964).
Maronea:	Magnusson (1934).
Mazosia:	Santesson (1952).
Medusulina:	Redinger (1933).
Megalospora:	Malme (1923).
Melanelia:	Esslinger (1977, 1978a).
Melanotheca:	Keissler (1937b); Malme (1924); Muller Argoviensis (1895).
Melaspilea:	Redinger (1936b); Wirth and Hale (1963).
Menegazzia:	Santesson (1943).
Micarea:	James and Coppins (1979); Vezda and Wirth (1976).
Microthelia:	Keissler (1937a); Muller Argoviensis (1895).
Mycoporellum:	Keissler (1937c).
Neofuscelia:	Esslinger (1977, 1978a).
Nephroma:	Wetmore (1960).
Neuropogon:	Lamb (1939).
Normandina:	Henssen (1976).

Ocellularia:	Hale (1978).
Ochrolechia:	Howard (1970); Verseghy (1962).
Opegrapha:	Hayward (1977); Redinger (1936b, 1940); Santesson (1952).
Pannaria:	Jorgensen 1978; Tavares (1965).
Pannoparmelia:	Bratt *et al.* (1976); Galloway (1978).
Parathelium:	Malme (1924).
Parmelia:	Hale (1965); Hale and Kurokawa (1964).
Parmeliella:	Dodge (1933); Malme (1925); Zahlbruckner (1941).
Parmelina:	Hale (1976c).
Parmeliopsis:	Hillman (1936).
Parmentaria:	Malme (1924).
Parmotrema:	Hale (1965b, 1974a).
Peltigera:	Kurokawa *et al.* (1966); Murray (1960); Thomson (1950).
Peltula:	Wetmore (1970).
Pertusaria:	Dibben 1980; Erichsen (1935); Malme (1936b); Oshio (1968).
Phaeographina:	See *Graphina.*
Phaeographis:	See *Graphina.*
Phaeophyscia:	Esslinger (1978b); Moberg (1977).
Phaeotrema:	Hale (1978)
Physcia:	Moberg (1977); Thomson (1963).
Physciopsis:	Moberg (1977); Poelt (1965); Thomson (1963).
Physma:	Hue (1905); Zahlbruckner (1941).
Pilophorus:	Jahns (1970).
Placopsis:	Lamb (1947).
Placynthium:	Henssen (1963b); Weber (1977).
Pleurotrema:	Malme (1924).
Polyblastiopsis:	Keissler (1937a).
Porina:	Malme (1929); Santesson (1952); Swinscow (1962).
Porocyphus:	Henssen (1963a).
Pseudevernia:	Hale (1968).
Pseudocyphellaria:	Magnusson (1940); Zahlbruckner (1941).
Pseudoparmelia:	Hale (1976a).
Psoroma:	Malme (1925); Zahlbruckner (1941).
Psoromaria:	Hue (1912).
Pyrenula:	Keissler (1937a).
Pyxine:	Imshaug (1957); Kashiwadani (1977); Swinscow and Krog (1975a).
Ramalina:	Krog and James (1977); Krog and Swinscow (1976a).
Relicina:	Hale (1975b).
Rhizocarpon:	Malme (1923); Runemark (1956).
Rinodina:	Sheard (1967).
Rocella:	Darbishire (1898).
Sagenidium:	Zahlbruckner (1941).
Sagiolechia:	Vezda (1968).
Sarcographa:	Redinger (1933).
Sarcogyne:	Magnusson (1935b).
Siphula:	Mathey (1971).
Sporastatia:	Magnusson (1935a).
Sporopodium:	Santesson (1952).
Spilonema:	Henssen (1963a).
Stenocybe:	Keissler (1937d).
Stereocaulon:	Galloway *et al.* (1976); Lamb (1977, 1978).

Sticta:	Zahlbruckner (1941).
Strigula:	Santesson (1952).
Tapellaria:	Santesson (1952).
Teloschistes:	Filson (1969).
Thamnolia:	Filson (1972).
Thelidium:	Zschacke (1933a).
Thelotrema:	Hale (1978).
Thysanothecium:	Abbayes (1968); Galloway (1977).
Tomasellia:	Keissler (1937b); Malme (1924).
Trapelia:	Hertel (1969a).
Tricharia:	Santesson (1952).
Trichothelium:	Santesson (1952).
Trimmatothele:	Zschacke (1933a).
Trypethelium:	Keissler (1937b); Malme (1924).
Umbilicaria:	Blackman *et al.* (1974); Kershaw (1961); Llano (1950, 1965).
Usnea:	Motyka (1936-38); Swinscow and Krog (1974, 1975b, 1978).
Verrucaria:	Swinscow (1968); Zschake (1933a).
Xanthoparmelia:	Hale (1974b).
Xanthoria:	Filson (1969).

References

Abbayes, H. des (1968) "Les lichens du genre *Thysanothecium* Mont. et Berk. (Cladoniacees) et leur constitution chimique", *J.Jap.Bot.* 43: 10-12.

Acharius, E. (1974) "Forsuk till en forbattrada Lafvarnes indelning (Dianome Lichenum)", *K. svensk Vetensk Akad. Handl.* 15:237ff. (not seen).

Acharius, E. (1803) *Methodus Quo Omnes Detectos Lichenes* (Ulrich, Stockholm).

Acharius, E. (1806) "*Arthonia*, novum genus Lichenum", *Neues J. Bot.* 1(3): 1-23.

Acharius, E. (1808) "Fortekning pa de in Sverige vaxtande arter af Lafvarnes familj", *K. svensk Vetensk Akad. Handl.* 29: 125, 228, 259 (not seen).

Acharius, E. (1810) *Lichenographia Universalis* (Dankwerts, Gottingen).

Acharius, E. (1814) *Synopsis Methodica Lichenum* (Svanborg et Soc., Lund).

Acharius, E. (1815) "Afhandling om de Cyrptogamiske vexter som komma under namn of Calicioidea," *K. svensk Vetensk Akad. Handl.* 37: 106-25, 246-91.

Adanson, M. (1763) *Familles des plantes*, 2 vols (Vincent, Paris).

Agardh, C.A. (1820-22) *Species algarum rite cognitae, cum synonymis, differentiis specificis et descriptionibus succinctis*, vol. 1 (Mauritius, Gryphiswald).

Ahti, T. (1961) "Taxonomic studies on the reindeer lichens (*Cladonia* subgenus *Cladina*)", *Suomal. elain-ja kasvit. Seur. van Julk.* 32(1): 1-160.

Almborn, O. (1966) "Revision of some Lichen genera in Southern Africa 1", *Bot. Notiser* 119: 70-112.

Almquist, S. (1880) "Monographia Arthoniarum Scandinaviae", *K. svensk Vetensk Akad. Handl.* 17(6) (not seen).

Alon, G. and Galun, M. (1971) "The genus *Caloplaca* in Israel", *Israel J. Bot.* 20: 273-92.

Arvidsson, L. and Galloway, D.J. (1979) "The lichen genus *Coccocarpia* in New Zealand", *Bot. Notiser* 132: 239-46.

Awasthi, D.D. (1975) "A monograph of the lichen genus *Dirinaria*", *Biblthca Lichenol.* 2: 1-108.

Awasthi, D.D. and Akhtar, P. (1977) "The genus *Leptogium* (sect. *Mallotium*) in India", *Norw. J. Bot.* 24: 59-71.

Bibby, P.N.S. (1955) "A remarkable lichen from arid Australia", *Muelleria* 1: 60.

Bird, C.D. (1974) "Studies on the lichen genus *Evernia* in North America", *Can. J. Bot.* 52: 2427-34.

Blackman, A.J., Bratt, G.C. and Cashin, J.A. (1974) "Umbilicariaceae in Tasmania", *Lichenologist* 6: 112-14.

Bornet, E. (156) "Descriptions de trois Lichens nouveaux", *Mem. Soc. Sci. Nat. Cherbourg* 4: 225-28.

Bory de St. Vincent, J.B.M. (1822-31) *Dictionnaire classique d'histoire naturelle* (Edit.) 17 vols (Rignoux, Paris).

Bowler, P.A. and Rundel, P.W. (1977) "Synopsis of a new lichen genus, *Fistulariella*", *Mycotaxon* 6: 195-202.

Bratt, G.C., Blackman, A.J. and Cashin, J.A. (1976) "The genus *Anzia* in Tasmania", *Lichenologist* 8: 69-77.

Brodo, I.M. (1973) "The lichen genus *Coccotrema* in North America", *Bryologist* 76: 260-70.

Brodo, I.M. and Hawksworth, D.L. (1977) "*Alectoria* and allied genera in North America", *Op. bot. Soc. bot. Lund* 42:1-164.

Choisy, M. (1949) "Catalogue des lichens de la region Lyonnaise", *Bull. mens. Soc. linn. Lyon* 18: 104-120, 137-52.

Choisy, M (1950) "Catalogue des lichens de la region Lyonnaise", *Bull. mens. Soc. linn. Lyon* 19: 9-24.

Crombie, J.M. (1879) "Enumeration of Australian lichens in herb. Robert Brown (Brit. Mus.) with descriptions of new species", *J. Linn. Soc. Bot.* 17: 390-401.

Culberson, C.F. (1969) *Chemical and Botanical Guide to Lichen Products* (University of North Carolina Press, Chapel Hill).

Culberson, C.F. (1970) "Supplement to *Chemical and Botanical Guide to Lichen Products*", *Bryologist* 73: 177-377.

Culberson, C.F., Culberson, W.L. and Johnson, A. (1977) *Second Supplement to Chemical and Botanical Guide to Lichen Products* (American Byological and Lichenological Society, St. Louis).
Culberson, W.L. (1964) "A summary of the lichen genus *Haematomma* in North America", *Bryologist* 66: 224-36.
Culberson, W.L. and Culberson, C.F. (1968) "The lichen genera *Cetrelia* and *Platismatia* (Parmeliaceae)", *Contr. U.S. natn. Herb.* 34: 449-558.
Darbishire, O.V. (1898) "Monographia *Rocelleorum*", *Biblthca. bot.* 45: 1-103.
Degelius, G. (1954) "The genus *Collema* in Europe", *Symb. bot. upsal. 13(2): 1-499.*
Degelius, G. (1974) "The lichen genus *Collema* with special reference to the extra-European species", *Symb. bot. upsal.* 20(2) 1-215.
Dibben, M.J. (1980) *The Chemosystematics of the Lichen Genus Pertusaria in North America North of Mexico.* (Milwaukee Museum, Milwaukee).
Dodge, C.W. (1933) "The foliose and fruticose lichens of Costa Rica I", *Ann. Mo. bot. Gdn.* 20: 373-467.
Elix, J.A. (1980) "A taxonomic revision of the lichen genus *Hypogymnia* in Australasia", *Brunonia* 3: 175-245.
Eschweiler, F.G. (1824) *Systema Lichenum, genera exhibens rite distincta, pluribus novis adaucta* (Schrag, Nuremburg).
Esslinger, T.L. (1973) "Chemical and taxonomic studies on some corticolous members of the lichen genus *Cetraria* in western North America", *Mycologia* 65: 602-13.
Esslinger, T.L. (1977) "A chemosystematic revision of the brown *Parmeliae*", *J. Hattori Bot. Lab.* 42: 1-211.
Esslinger, T.L. (1978a) "A new status for the brown *Parmeliae*", *Mycotaxon* 7: 45-54.
Esslinger, T.L. (1978b) "Studies in the lichen family Physciaceae II. The genus *Phaeophyscia* in North America", *Mycotaxon* 7: 283-320.
Fee A.L.A. (1824-25). *Essai sur les Crypotogames des ecorces exotiques officinales* (Firmin-Didot, Paris).
Fee, A.L.A. (1837) *Essai sur les Cryptogames des ecorces exotiques officinales. II. Supplement et revision* (Firmin-Didot, Paris).
Filson, R.B. (1969) "A review of the genera *Teloschistes* and *Xanthoria* in the lichen family Teloschistaceae in Australia", *Muelleria* 2: 65-115.
Filson, R.B. (1970) "Studies in Australian Lichens 1", *Vict. Nat.* 87: 324-27.
Filson, R.B. (1972) "Studies in Australian Lichens II. The alpine lichen *Thamnolia vermicularis* (Sw.) Schaer. in Australia", *Muelleria* 2: 180-87.
Filson, R.B. (1978) "A revision of the genus *Heterodea* Nyl.", *Lichenologist* 10: 13-25.
Filson, R.B. and Rogers, R.W. (1979) *The Lichens of South Australia* (Government Printer, Adelaide).
Flotow, J. (1851) "Briefe uber *Psora privigna* (Ach.) Fw. 1848. (*Sarcogyne corrugata* Fw. ad int. 1845, *Lichen simplex* Dav., *Patellaria simplex* Wallr., *Lecidea* Hook)", *Bot. Ztg.* 9: 753-59, 769-76.
Forssell, K.B.J. (1885) "Beitrage zur Kentniss der Anatomie u. Systematik der Gloeolichenen", *Nova Acta R. Soc. Scient. upsal.* ser. 3, 13: 1-118.
Frey, E. (1933) "*Icmadophila*", *Rabenh. Krypt.-Fl. 9*, 4(1): 19-25.
Fries, E.M. (1821) "Beskrifning pa Nya Lafslagten", *K. svensk Vetensk Akad. Handl.* 1821: 321-24.
Fries, E.M. (1825) *Systema orbis vegetabilis. Primas lineas novae constructionis periclitatur Elias Fries. Pars 1. Plantae homoemeae* (Typographica Academica, Lund).
Fries, E.M. (1846) *Summa vegetabilium Scandinaviae* (Bonnier: Stockholm).
Fries, T.M. (1857) *De Stereocaulis et Pilophoris commentatio* (Wahlstrom, Uppsala).
Fries, T.M. (1861) "Lichenes arctoi Europaeae Gronlandiaeque hactenus cogniti", *Nova Acta R. Soc. Scient. upsal.* (3)3: 103-318.
Fries, T.M. (1871) *Lichenographia Scandinavica sive dispositio lichenum in Dania, Suecia, Norvegia, Fennia, Lapponia, Rossica hactenus collectorum. Pars Prima* (E. Berling, Uppsala).
Fries, T.M. (1874) *Lichenographia Scandinavica sive disposito lichenum in Dania, Suecia, Norvegia, Fennia, Lapponia, Rossica hactenus collectorum. Pars Secunda* (E. Berling, Uppsala).
Galloway, D.J. (1977) "The lichen genus *Thysanothecium* Mont. and Berk., and historical note", *Nova Hedwigia* 28: 499-512.
Galloway, D.J. (1978) "*Anzia* and *Pannoparmelia* (Lichenes) in New Zealand", *N.Z. J. Bot.* 16: 261-70.
Galloway, D.J., Lamb, I.M. and Bratt, G.C. (1976) "Two new species of *Stereocaulon* from New Zealand and Tasmania", *Lichenologist* 8: 61-67.
Gray, S.F. (1821) *A natural arrangement of British Plants according to their relationships to each other . . ., etc.* 2 vols. (Baldwin, Cradock and Joy, London).
Hakulinen, R. (1854) "Die flechtengattung *Candelariella* Muller Argoviensis", *Suomal. elain-ja kasvit. Seur. van. Julk.* 27(3): 1-127.
Hale, M.E. (1965a) "Studies on the *Parmelia borreri* group", *Svensk. Bot. Tidskr.* 59: 37-48.
Hale, M.E. (1965b) "A monograph of *Parmelia* subgenus *Amphigymnia*", *Contr. U.S. natn. Herb.* 36: 193-358.
Hale, M.E. (1967) *The Biology of Lichens* (Edward Arnold, London).
Hale, M.E. (1968) "A sysnopsis of the lichen genus *Pseudevernia*", *Bryologist* 71: 1-11.

Hale, M.E. (1974a) "New combinations in the lichen genus *Parmotrema* Massalongo", *Phytologia* 28: 334-49.
Hale, M.E. (1974b) "Delimitation of the Lichen Genus *Hypotrachyna* (Vainio) Hale", *Phytologia* 28: 340-42.
Hale, M.E. (1974c) "*Bulbothrix, Parmelina, Relicina* and *Xanthorparmelia,* four new genera in the Parmeliaceae (Lichenes)", *Phytologia* 28: 249-90.
Hale, M.E. (1975a) "A revision of the lichen genus *Hypotrachyna* (Parmeliaceae) in tropical America", *Smithson. Contr. Bot.* 25: 1-73.
Hale, M.E. (1975b) "A monograph of the lichen genus *Relicina* (Parmeliaceae)", *Smithson. Contr. Bot.* 26: 1-32.
Hale, M.E. (1976a) "A monograph of the lichen genus *Pseudoparmelia* Lynge (Parmeliaceae)", *Smithson. Contr. Bot. 32: 1-29.*
Hale, M.E. (1976b) "A monograph of the lichen genus *Bulbothrix* Hale (Parmeliaceae)", *Smithson. Contr. Bot.* 32: 1-29.
Hale, M.E. (1976c) "A monograph of the lichen genus *Parmelina* Hale (Parmeliaceae)", *Smithson. Contr. Bot.* 33: 1-60.
Hale, M.E. (1978) "A revision of the lichen family Thelotremaceae in Panama", *Smithson. Contr. Bot.* 38: 1-60.
Hale, M.E. and Kurokawa, S. (1964) "Studies on *Parmelia* subgenus *Parmelia*", *Contr. U.S. natn. Herb.* 36:121-91.
Hawksworth, D.L., James, P.W. and Coppins, B.J. (1980). Checklist of British lichen-forming, lichenicolous and allied fungi. *Lichenol.* 12, 1-115.
Hayward, G.C. (1977) "Taxonomy of the lichen families Graphidaceae and Opegraphaceae in New Zealand", *N.Z. J. Bot.* 15: 565-84.
Hedlund, T. (1892) "Kritische bemerkungen uber einige Arten der Flechtengattung *Lecanora* Ach. *Lecidea* Ach. and *Micarea*", *K. svensk Vetensk Akad. Handl.* 18(3)3: 1-104.
Hedwig, J. (1787-1797). *Descriptio et adumbratio microscopica-analytica Muscorum frondoscorum nec non aliorum vegetantium e classe crypotgamica Linnaei novorum dubiisque Vexatorum. Stirpium cryptogamicarum* (J.G. Muller, Lipsiae).
Henssen, A. (1963a) "Eine Revision der Flechtenfamilien Lichenaceae und Ephebaceae", *Symb. bot. upsal.* 18(1): 1-123.
Henssen, A. (1963b) "The North American species of *Placynthium*", *Can. J. Bot.* 41: 1687-1724.
Henssen, A. (1969) "Eine studie uber die Gattung *Arctomia*", *Svensk Bot. Tidskr.* 63: 126-38.
Henssen, A. (1976) "Studies in the developmental morphology of lichenised ascomycetes", in *Lichenology: Progress and Problems* ed. D.H. Brown, D.L. Hawksworth and R.H. Bailey (Academic Press, London).
Hepp, P. (1853-67). *Die Flecten Europas in getrockneten mikroskopisch untersuchten Exemplaren mit Beschreibung v. Abbildung ihrer Sporen* 16 Fasc. (Zurich).
Hertel, H. (1967) "Revision einiger Calciphiler formenkreise der Flechtengattung *Lecidea*", *Beih. Nova Hedwigia* 24: 1-155.
Hertel, H. (1968-75) "Beitrage zur Kenntnis der Flechtenfamilie Lecideaceae I-VI", *Herzogia* 1: 25-39, 321-29; 2: 37-62, 231-61, 479-515; 3: 365-406.
Hertel, H. (1969a) "Die flechtengattung *Trapelia* Choisy", *Herzogia* 1: 111-30.
Hill, J. (1773) *A General Natural History, Vol. II: A History of Plants.* Edit. 2 (Osborne, London).
Hillman, J. (1936) "*Parmeliopsis*", *Rabenh. Krypt.-Fl.* 9 (5)1: 25-42.
Hoffman, G.F. (1789-1801) *Descriptio et adumbratio Plantarum e classe cryptogamica Linnaei, quae Lichenes dicuntur* (Crasium, Lipsiae).
Hooker, J.D. (1847) "The Botany of the Antartic Voyage of H.M. discovery ships *Erebus & Terror*", (London).
Howard, G.E. (1970) "The lichen genus *Ochrolechia* in North America north of Mexico", *Bryologist* 73: 93-130.
Hue, A. (1905) "*Physma* unum e familiae Collemacearum generibus morphologice et anatomice", *Bull. Soc. Linn. Normandie* (5)9: 119-37.
Hue, A.M. (1908a) "Lichens" in *Expedition antarctique francaise 1903/05, commandee par le Dr. Jean Charcot. Botanique* (Masson, Paris). (not seen).
Hue, A. (1908b) "Monographia generis *Solorinae* Ach. morphologice et anatomice, addito de genera *Psoromaria* Nyl. appendice", *Mem. Soc. Sci. Nat. Math. Cherbourg.* 38: 1-56.
Hue, A. (1924) "Monographia *Crocyniarum*", *Bull. Soc. Bot. Fr.* 71: 311-402.
Humboldt, F.A. (1793). *Florae Fribergensis specimen plantas cryptogamicas praesertim subterranes exhibens* (Rottmann, Berlin).
Imshaug, H.A. (1955) "The lichen genus *Buellia* in the West Indies", *Farlowia* 4: 473-512.
Imshaug, H.A. (1957) "The lichen genus *Pyxine* in North and Middle America", *Trans. Am. Micros. Soc.* 76: 246-69.
Imshaug, H.A. and Brodo, I.M. (1966) "Biosystematic studies on *Lecanora pallida* and some related lichens in the Americas", *Nova Hedwigia* 12: 1-59.

Jahns, H.M. (1970) "Remarks on the taxonomy of the European and North American species of *Pilophorus* Th.Fr.", *LIchenologist* 4: 199-213.
James, P.W. (1973) "New or interesting British lichens", *Lichonologist* 5: 114-48.
Jordan, W.P. (1973) "The genus *Lobaria* in North America north of Mexico", *Bryologist* 76: 225-51.
Jorgensen, P.M. (1972) "Further studies in *Alectoria* sect. *Divaricatae*", *Svensk. bot. Tidskr.* 66: 191-201.
Jorgensen, P.M. (1975) "Contributions to a monograph of the *Mallotium*-hairy *Leptogium* species", *Herzogia* 3: 433-60.
Jorgensen, P.M. (1978) "The lichen family Pannariaceae in Europe". *Op. bot. Soc. bot. Lund* 45: 1-123.
Karnefelt, I (1977) "Three new species of brown fruticose *Cetraria*", *Bot. Notiser* 130: 125-29.
Kashiwadani, H. (1977) "The genus *Pyxine* (Lichens) in Papua-new Guinea", *Bull. natn. Sci. Mus. Tokyo Ser. B. (Bot.)* 3(3): 63-70.
Keissler, K. (1937a) "Familie Pyrenulaceae", *Rabenh. Krypt.-Fl.* 9, 1(2): 1-419.
Keissler, K. (1937b) "Familie Trypetheliaceae", *Rabenh. Krypt.-Fl.* 9, 1(2): 422-39.
Keissler, K. (1937c) "Familie Mycoporaceae", *Rabenh. Krypt.-Fl.* 9, 1(2): 470-502.
Keissler, K. (1937d) "Caliciaceae", *Rabenh. Krypt.-Fl.* 9, 1(2): 520-714.
Kershaw, K.A. (1961) "The genus *Umbilicaria* in the British Isles", *Lichenologist* 1: 251-65.
Keuck, G. (1977) "Ontogenetisch-systematische Studie uber *Erioderma* in Vergleich mit anderen cyanophilen Flechtengattungen", *Biblthca. Lichenol.* 6: 1-175.
Korber, G.W. (1848) *Grundriss der Krypotgamen-Kunde. Zur Orientirung beim Studium der Kryptogamischen Pflanzen, sowie zum Gebrauch bei seinen Vorlesungen* (E. Trewendt, Breslau).
Korber, G.W. (1855) *Systema lichenum Germaniae* (Trewendt and Granier, Breslau).
Krempelhuber, A. (1876) "Aufzahlung und Beschreibung der Flechtenarten, welche Dr. Heinrich Wawra Ritter von zwei Reisen um die Erde mitbrachte", *Verh. zool.-bot. Ges. Wein* 26: 433-46.
Krog, H. and James, P.W. (1977) "The genus *Ramalina* in Fennoscandia and the British Isles", *Norw. J. Bot.* 24: 15-43.
Krog, H. and Swinscow, T.D.V. (1976a) "The genus *Ramalina* in East Africa", *Norw. J. Bot.* 23; 153-75.
Krog, H. and Swinscow, T.D.V. (1976b) "*Ramalina* species with a hollow thallus *(Fistularia)* in East Africa", *Norw. J. Bot.* 21: 111-24.
Krog, H. and Swinscow, T.D.V. (1979) "*Parmelia* subgenus *Hypotrachyna* in East Africa", *Norw. J. Bot.* 26: 11-43.
Kurokawa, S. (1962) "A monograph of the genus *Anaptychia*", *Beih. Nova Hedwigia* 6: 1-115.
Kurokawa, S. (1973) "Supplementary notes on the genus *Anaptychia*", *J. Hattori bot. Lab.* 37: 563-607.
Kurokawa, S., Jenzenji, Y., Shibata, S. and Chiang, H.C. (1966) "Chemistry of Japanese *Peltigera* with some taxonomic notes", *Bull. Natn. Sci. Mus. Tokyo* 9(2): 101-14.
Kurokawa, S. and Kashiwadani, H. (1977) "Notes on the lichen genus *Lopadium* in Japan and Formosa", *Bull. Natn. Sci. Mus. Tokyo* 3(4): 124-34.
Lamarck, J.B.P.A. de M. De and Candolle, A.P. de (1805-1815) *Flore Francaise, ou descriptions succinctes de toutes plantes qui croissent naturellement en France . . ., etc.* 5 vols. (Desray, Paris).
Lamb, I.M. (1939) "A review of the genus *Neuropogon* (Nees & Flot.) Nyl., with special reference to the Antarctic species", *J. Linn. Soc., Bot.* 52: 199-237.
Lamb, I.M. (1947) "A monograph of the lichen genus *Placopsis* Nyl.", *Lilloa* 13: 151-288.
Lamb, I.M. (1977) "A conspectus of the lichen genus *Stereocaulon* (Schreb.) Hoffm.", *J. Hattori Bot. Lab.* 43:191-355.
Lamb, I.M. (1978) "Keys to the species of the lichen genus *Stereocaulon* (Schreb.) Hoffm.", *J. Hattori Bot. Lab.* 44: 209-50.
Lamb, I.M. and Ward, A. (1974) "A preliminary conspectus of the species attributed to the imperfect lichen genus *Leprocaulon* Nyl.", *J. Hattori bot. Lab.* 38: 499-553.
Lamy, De La Chapelle, E. (1868) "Catalogue raissone des lichens du Mont-Dore et de al Haut-Vienne", *Bull. Soc. Bot. Fr.* 25: 321-536.
Lettau, G. (1937) "Monographische Bearbeitung einiger Flechtenfamilien 1. Lecanactidaceae", *Reprium nov. Spec. Regni veg. Beih.* 69(2): 20-71.
Lindsay, L.W. (1866) "Observations on New Zealand Lichens", *Trans. Linn. Soc.* 25: 493-560.
Link, H.F. (1833) *Grundlehren der Krauterkunde* (Berlin) (not seen).
Llano, G.A. (1950) *A monograph of the lichen family Umbilicariaceae in the Western Hemisphere* (Navexos, Washington D.C.).
Llano, G.A. (1965) "Review of *Umbilicaria* Hoffm. and of the *Lasallias*", *Hvalrad Skr.* 48: 112-24.
Lynge, B. (1914) "Die flechten der ersten Regnellschen Expedition. Die gattungen *Pseudoparmelia* gen.nov. und *Parmelia* Ach.", *Ark. Bot.* 13(13): 1-170.
Magnusson, A.H. (1929) "A monograph of the genus *Acarospora*", *K. svensk Vetensk Akad. Handl.* Ser. 3, 7(4): 1-400.
Magnusson, A.H. (1934) "Die flechtengattung *Maronea* Mass.", *Acta Horti gothoburg* 9: 41-66.
Magnusson, A.H. (1935a) "*Sporastatia*", *Rabenh. Krypt.-Fl.* 9, 5(1): 1-15.
Magnusson, A.H. (1935b) "*Sarcogyne*", *Rabenh. Krypt.-Fl.* 9, 5(1): 49-104.
Magnusson, A.H. (1935c) "*Biatorella*", *Rabenh. Krypt.-Fl.* 9, 5(1): 15-46.

Magnusson, A.H. (1940) "Lichens from Central Asia 1", *Rep. Sino-Swedish sci. Exp. N.W. China. XI (Bot).* 1: 1-167.
Magnusson, A.H. (1940) "Studies in species of *Pseudocyphellaria.* The *Crocata* group", *Göteborgs Bot. Träd* 14, 1-36.
Magnusson, A.H. (1953) "Key to saxicolous *Buellia* species, mainly from South America", *Ark. Bot.* 3(9): 205-21.
Magnusson, A.H. (1955) "A catalogue of Hawaiian lichens", *Ark. Bot.* 3(10): 223-402.
Malme, G.A. (1923) "Die flechten der ersten Regnellschen Expedition. Die gattungen *Bombyliospora, Megalospora, Catillaria* und *Rhizocarpon*", *Ark. Bot.* 18(12): 1-17.
Malme, G.O. (1925) "Die Pannariazeen der Regnellschen Herbars.", *Ark. Bot.* 20A(3): 1-23.
Malme, G.A. (1926) "Lichenes Blasteniospori herbarii Regnelliani", *Ark. Bot.* 20A(9): 1-51.
Malme, G.A. (1928) "*Buelliae* itineris Regnelliani primi", *Ark. Bot.* 21A(14): 1-42.
Malme, G.O. (1929) "Lichenes Itinere Regnelliano primo collecti", *Ark. Bot.* 23A: 1-37.
Malme, G.O. (1934) "Die Gyalectazeen der ersten Regnellschen Expedition", *Ark. Bot.* 26A(13): 1-10.
Malme, G.O. (1935) "*Bacidiae* intineris Regnelliani primi", *Ark. Bot.* 27A(5): 1-40.
Malme, G.O. (1936) "*Pertusariae* Expeditionis Regnellianae primae", *Ark. Bot.* 28A(9): 1-27.
Martin, W. (1958) "The *Cladoniae* of New Zealand", *Trans. R. Soc. N.Z.* 85: 603-32.
Martin, W. (1965) "The lichen genus *Cladia*", *Trans. R. Soc. N.Z.* 3: 7-12.
Massalongo, A.B. (1852) *Ricerche sull'autonomia dei Licheni crostosi e Materiali pella loro naturale ordinazoine* (Frizerio, Verona).
Massalongo, A.B. (1853) *Alcuni generi de Licheni nouvamente limitati e descritti* (Antonelli, Verona).
Massalongo, A.B. (1854a) *Neagenea Lichenum* (Ramanzini, Verona).
Massalongo, A.B. (1854b) *Geneacaena Lichenum noviter proposita ac descripta* (Verona) (not seen).
Massalongo, A.B. (1855) *Frammenti Lichenographica* (Verona).
Massalongo, A.B. (1856a) "De nonnullis Collemaceis ex tribu Omphulariearum brevis commentatio", *Flora, Jena* 39: 209-15.
Massalongo, A.B. (1856b) *Miscellanea lichenologica* (Civelli: Verona-Milano).
Massalongo, A.B. (1860) "Esame Comparativo di alcune generi de lichini", *Atti Ist. veneto Sci. ser. 3,* 5: 247-76, 313-37.
Mathey, A. (1971) Contribution a l'etude du genre *Siphula* (Lichens) en Afrique. *Nova Hedwigia* 22, 795-878.
Meyen, F.J.F. and Flotow, J. (1843) "Observationes in itinere circum terram institutae (1830-1832): Lichenes", *Nov. Acta Acad. Caesar. Leop. Carol.* 19 Suppl. 209-32.
Moberg, R. (1977) "The lichen genus *Physcia* and allied genera in Fennoscandia", *Symb. bot. upsal.* 22(1): 1-108.
Montagne, J.F.C. (1851) "Cryptogamia Guyanensis seu plantarum cellularium in Guyana gallica annis 1835-49 a cl. Leprieur collectorum enumeratio universalis", *Ann. Sci. Nat. Bot.* Ser. 3, 16: 29-63, 91-144, 311-29.
Montagne, J.F.C. (1852) Diagnose phycologicae seu quibus caracteribus discriminandae sunt species lichenum algarumque nonnullae novae in tomo florae Chilensis octavo nondum typis mandato descriptae. *Annls. Sci. nat. Bot.* Ser. 3, 18, 302-319. (not seen).
Montagne, J.F.C. and Berkeley, M.J. (1846) "On *Thysanothecium*, a new genus of Lichens", *London J. Bot.* 5: 257-58.
Montagne, J.F.C. and van den Bosch, R.B. (1855) "Lichenes" in F.W. Junghuhn, *Plantae Junghuhnianae* (Lugduni, Batavia) (not seen).
Motyka, J. (1936-38) *Lichenum generis Usea studium Monographicum, pars systematica,* 2 vols (Motyka, Leipzig).
Motyka, J. (1964) "The North America species of Alectoria", *Bryologist* 67: 1-44.
Muller Argoviensis, J. (1862) "Principes de Classification des Lichens et Enumeration des Lichens de Geneve", *Mem. Soc. Phys. Hist. nat. Geneve* 16(2): 343-433.
Muller Argoviensis, J. (1880a) "Lichenologische Beitrage X", *Flora, Jena* 63: 17-24, 40-45.
Muller Argoviensis, J. (1880b) "Lichenologische Beitrage XI", *Flora, Jena* 63: 259-68, 275-90.
Muller Argoviensis, J. (1881) "Lichenologische Beitrage XIV", *Flora, Jena* 64: 505-11, 513-27.
Muller Argoviensis, J. (1882) "Lichenologische Beitrage XVII", *Flora, Jena* 65: 483-90, 499-505, 515-19.
Muller Argoviensis, J. (1883) "Lichenologische Beitrage XVIII", *Flora, Jena* 66: 243-49, 271-74, 286-90, 304-6, 317-22, 330-8, 344-54.
Muller Argoviensis, J. (1884) "Lichens de Palestine", *Rev. Mycol.* 6: 12-15.
Muller Argoviensis, J. (1885) "Pyrenocarpeae cubenses", *Bot. Jb.* 6: 374-421.
Muller Argoviensis, J. (1887) "Lichenologische Beitrage XXV", *Flora, Jena* 70: 56-64, 74-80.
Muller Argoviensis, J. (1887a) "Lichenologische Beitrage XXVI", *Flora, Jena* 70: 268-73, 283-88, 316-22, 326-38, 396-402, 423-29.
Muller Argoviensis, J. (1887b) "Graphideae Feeanae inclus. trib. affinibus nec non Graphideae exoticae Acharii, El. Friesii et Zenkeri de novo studio spiciminum originalum expositae et in novam dispositionem ordinatae", *Mem. Soc. Phys. Hist. nat. Geneve* 29(8): 1-80.

Muller Argoviensis, J. (1889) "Lichens" in *Mission Scientifique du Cap Horn vol. 5 Botanique* (Paris).
Muller Argoviensis, J. (1891) "Lichenes Brisbanensis", *Nuovo Giorn. Bot. Ital.* 23: 385-404.
Muller Argoviensis, J. (1892) "Lichenes Manipurensis", *J. Linn. Soc. Bot.* 29: 217-31.
Muller Argoviensis, J. (1893) "Lichenes Wilsonianae", *Bull. Herb. Boissier* 1: 33-65.
Muller Argoviensis, J. (1894a) "Lichenes Eckfeldtiani", *Bull. Herb. Boissier* 2: 89-93.
Muller Argoviensis, J. (1894b) "Conspectus systematicus lichenum Novae Zelandiae quem elaboravit", *Bull. Herb. Boissier* 2, append. 1: 1-114.
Muller Argoviensis, J. (1895) "Pyrenocarpeae Queenslandiae", *Australasian Assoc. Adv. Sci.* 1895: 1-18.
Murray, J. (1960) "Studies of New Zealand Lichens 3. Peltigeraceae", *Trans. R. Soc. N.Z. Bot.* 88: 381-99.
Nakanishi, M. (1966) "Taxonomical studies on the family Graphidaceae of Japan", *J. Sci. Hiroshima Univ. Ser. B. div. 2 (bot.)* 11: 51-126.
Nees v. Esenbeck, C.G. (1820) *Horae Physicae berolinenses collectae ex Symbolis viroorum doctorum* (Bonn).
Norman, J.M. (1853) "Conatus praemissus redactionis novae generum nonnullorum Lichenum in organis fructificationis vel sporis fundatae", *Nytt Mag. Naturv.* 7: 213-52.
Notaris, de G. (1846) "Frammenti lichenografici di un lavora inedito", *Giorn. Bot. Ital.* 2(1): 174-224, 299-330.
Nylander, W. (1853) "Lichenes Algeriensis novi", *Annls. Sci. nat. Bot. ser.* 3, 20: 314-20.
Nylander, W. (1854) "De nyaste botaniska undersokningarne af mjoldrygan", *Bot. Notiser* 1854, p. 33 ff (not seen).
Nylander, W. (1855) "Essai d'une nouvelle classification des Lichens", *Mem. Soc. Sci. Nat. Cherbourg* 2: 5-16.
Nylander, W. (1856) "Prodromus Lichenographiae Galliae et Algeriae", *Act. Soc. Linn. Bordeaux* 21: 249-467.
Nylander, W. (1858-1860) *Synopsis methodica Lichenum*, vol. 1 (Martinet, Paris).
Nylander, W. (1862) "*Tylophoron* et *Parathelium*, genera Lichenum nova", *Bot. Ztg.* 20: 278-79 (not seen).
Nylander, W. (1886) "Collectio Lichenum ex Insula Cuba", *Flora, Jena* 49: 289-95.
Nylander, W. (1867) "Synopsis Lichenum Novae Caledoniae", *Bull. Soc. Linn. Normandie ser 2,* 2: 39-140.
Nylander, W. (1868) "Lichenes Angolenses Welsitschiani", *Bull. Soc. linn. Normandie ser 2,* 2: 508-21.
Nylander, W. (1870) "Recognitio monographica Ramalinarum", *Bull. Soc. Linn. Normandie ser 2,* 6: 256-328.
Nylander, W. (1888) *Lichenes Novae Zelandiae* (P. Schmidt, Paris).
Nylander, W. (1896) *Les Lichens des environs de Paris* (P. Schmidt, Paris).
Olivier, H. (1897) *Expose systematique et descriptions des Lichens de l'Ouest et du Nord-Ouest de la France (Normandie, Bretagne, Anjou, Maine, Vendee)* (Klinck sieck, Paris) (not seen).
Oshio, M (1968) "Taxonomical studies on the family Pertusariaceae of Japan", *J. Sci. Hiroshima Univ. ser. B. div. 2 (Botany)* 12: 81-163.
Parmasto, E. (1978) "The genus *Dictyonema* (Thelephorolichenes)", *Nova Hedwigia* 29: 99-144.
Persoon, C.H. (1794) "Einige Bemerkungen uber die Flechten, nebst Beschreibungen einiger neuen Arten aus dieser Familie der Aftermoose", *Ann. Bot. (Usteri)* 7: 1-32, 155-58.
Persoon, C.H. (1826) "Lichens" in Gaudichaud, *Botanique, Voyage Autour du monde, execute sur les corvettes de S.M. L'Uranie et la Physicienne pendant les annees 1817-1820,* 187-215 (Pillet Aine, Paris).
Poelt, J. (1952) "Die *Lecanora subfusca* — Gruppe in suddeutschland", *Ber. bayer. bot. Ges.* 29: 58-69.
Poelt, J. (1958) "Die lobaten Arten der Flechtengattung *Lecanora* Ach. *sensu ampl.* in der Holarktis", *Mitt. bot. Stamml. Munch.* 2: 411-589.
Poelt, J. (1965) "Uber einiger artengruppen der Flechtengattungen *Caloplaca* und *Fulgensia*", *Mitt. bot. Stamml. Munch.* 5: 571-607.
Poelt, J. (1965) "Zur systematik der Flechtenfamilie Physciaceae", *Nova Hedwigia* 9: 21-32.
Poelt, J. (1973) "Classification" in *The Lichens* (V. Ahmadjian and M.E. Hale Eds) (Academic Press, London).
Poelt, J. (1974) "Zur Kenntis der flechtenfamilie Candelariaceae", *Phyton (Austria)* 16: 189-210.
Redinger, K. (1933) "Die Graphidineen Flechten der ersten Regnellschen Expedition nach Brasilien 1892-94. I. *Glyphis, Medusulina* und *Sarcographa*", *Ark. Bot.* 25A(13): 1-20.
Redinger, K. (1936a) "Arthoniaceae", *Rabenh. Krypt.-Fl.* 9, 2(1): 1-180.
Redinger, K. (1936b) "Graphidaceae", *Rabenh. Krypt.-Fl.* 9, 2(1): 181-404.
Redinger, K. (1938) "Restitution und kritische Revision der Flechtengattung *Enterographa* Fee und *Sclerophyton* Esch.", *Bei. Feddes Report. Spec. Nov.* 43 (not seen).
Redinger, K. (1940) "Die Graphidineen der ersten Regnellschen Expedition nach Brasilien 1892-94. IV. *Opegrapha*", *Ark. Bot.* 29A (19): 1-52.
Reichenbach, H.G.L. (1841) *Der Deutsche Botaniker I. Das Herbarienbuch* (Dresden and Leipzig) (not seen).
Reidl, H. (1964) "Bemerkunger uber *Dermatina* — Arten aus West- und Mitteleuropa Vorlaufige Mitteilung", *Sydowia* 17: 102-13.
Runemark, H. (1956) "Studies in *Rhizocarpon* I. Taxonomy of the yellow species in Europe", *Op. bot. Soc. bot. Lund* 2(2): 1-150.

Santesson, R. (1943) "The South American *Menegazziae*", *Ark. Bot.* 30A(11): 1-35.
Santesson, R. (1952) "Foliicolous lichens I. A revision of the taxonomy of the obligately foliicolous lichenized fungi", *Symb. bot. Upsal.* 12: 1-590.
Schaerer, L.E. (1850) *Enumeratio critica lichenum europaeorum* (Bern) (not seen).
Schreber, J.C.D. (1791) in *C. Linnaeus Genera Planarum Ed. 8* (Magdeburg).
Sheard, J.W. (1964) "The genus *Buellia* de Notaris in the British Isles", *Lichenologist* 2: 225-62.
Sheard, J.W. (1967) "A revision of the lichen genus *Rinodina* (Ach.) Gray in the British Isles", *Lichenologist* 3: 328-67.
Sierk, H.A. (1964) "The genus *Leptogium* in North America north of Mexico", *Bryologist* 67: 245-317.
Sprengel, K.P.J. (1805) *Anleitung zur Kenntniss der Gewachse. III: Einleitung in das Studium der Cryptogamischen Gewachse* (Jummel, Halle).
Steiner, J. (1911) "Flechten aus den italienisch-franzosischen Grenzebiete und Mittelitalien", *Verh. zool.-bot. Ges. Wien* 61: 29-64.
Stirton, J. (1877) "Additions to the lichen flora of New Zealand and the Chatham Islands", *Proc. Phil. Soc. Glasgow* 10: 292-306.
Stizenberger, E. (1861) "*Anzia*, eine neue Flechtengattung", *Flora, Jena* 44: 390-93.
Swinscow, T.D.V. (1962) "Pyrenocarpous lichens: 3. The genus *Porina* in the British Isles", *Lichenologist* 2: 6-56.
Swinscow, T.D.V. (1968) "Pyrenocarpous lichens: 13. Fresh-water species of *Verrucaria* in the British Isles", *Lichenologist* 4: 34-54.
Swinscow, T.D.V. and Krog, H. (1974) "*Usnea* subgenus *Eumitria* in East Africa", *Norw. J. Bot.* 21: 165-85.
Swinscow, T.D.V. and Krog, H. (1975a) "The genus *Pyxine* in East Africa", *Norw. J. Bot.* 22: 43-68.
Swinscow, T.D.V. and Krog, H. (1975b) "The *Usnea undulata* aggregate in East Africa", *Lichenologist* 7: 121-38.
Swinscow, T.D.V. and Krog, H. (1975c) "The genus *Dermatocarpon* in East Africa with an overlooked species in Britain", *Lichenologist* 7: 148-54.
Swinscow, T.D.V. and Krog, H. (1976a) "The genera *Anaptychia* and *Heterodermia* in East Africa", *Lichenologist* 8: 103-38.
Swinscow, T.D.V. and Krog, H. (1976b) "The genus *Coccocarpia* in East Africa", *Norw. J. Bot.* 23: 251-59.
Swinscow, T.D.V. and Krog, H. (1978) "Pendulous species of *Usnea* in East Africa", *Norw. J. Bot.* 25: 221-41.
Tavares, C.N. (1965) "The genus *Pannaria* in Portugal", *Port. Acta biol. ser. B.* 8: 1-16.
Thomson, J.W. (1950) "The species of *Peltigera* of North America north of Mexico", *Am. Midl. Nat.* 44: 1-68.
Thomson, J.W. (1963) "The lichen genus *Physcia* in North America", *Beih. Nova Hedwigia* 7: 1-172.
Thomson, J.W. (1967) "The lichen genus *Baeomyces* in North America north of Mexico", *Bryologist* 70: 285-98.
Thomson, J.W. (1968) *The lichen genus Cladonia in North America* (University of Toronto Press, Toronto).
Thwaites, G.H.K. (1849) "Notes on Cystocoleus, a new genus of minute plants", *Ann. Nat. Hist.* 3: 241-42 (not seen).
Tibell, L. (1971) "The genus *Cyphelium* in Europe", *Svensk. Bot. Tidsk.* 65: 138-64.
Tibell, L. (1975) "The Caliciales of boreal North America", *Symb. bot. upsal.* 21(2): 1-136.
Tobler, F. (1937) "Uber den Bay der Hynmenolichenen und eine neue zu ihnen gehronde Gattung", *Flora (Jena),* 131, 438-47.
Trevisan, V. (1951-52) "Memoria sul Tetranychus Passerinii", *Riv. period Lav. I.R. Acad. Padua* 1851-52.
Trevisan, V. (1853) *Spighe e Paglie. Scritti botanici varii* (Luglio, Padua) (not seen).
Trevisan, V. (1868) "Sul genere *Dimelaena* di Norman", *Atti Soc. Ital. Sci. Nat. Milano* 11: 604-50.
Trevisan, V. (1880) "Sulle Garovaglinee, nuova tribu de Collemacee", *Rc. Ist. Lomb. Sci. Lett. ser. 2* 13: 65-77.
Tuckerman, E. (1877) "Observationes Lichenologicae No. 4. Observations on North American and other Lichens", *Proc. Am. Acad. Arts Sci.* 12: 106-85.
Vainio, E.A. (1890) "Etude sur la classification naturelle et la morphologie des Lichens du Bresil. Pars Prima", *Acta Soc. fauna flora fenn.* 7(1): 1-247.
Vainio, E.A. (1890) "Etude sur la classification naturelle et la morphologie de lichens du Bresil. Pars Secunda", *Acta. Soc. fauna flora fenn.* 7(2): 1-256.
Vainio, E.A. (1922) "Lichenographia Fennica II", *Acta Soc. fauna flora fenn.* 53(1): 1-340.
Verseghy, K. (1962) "Die Gattung *Ochrolechia*", *Beih. Nova Hedwigia* 1: 1-146.
Vezda, A. (1968) "Flechtensystematische Studien VI, Die Gattung *Sagiolechia* Massal.", *Folia geobot. phytotax. Praha* 2: 383-96.
Vezda, A. (1977) "Flechtensystematische studien X. *Conotremopsis weberiana gen. novum* et *sp. nova*, eine neue Flechte aus der Familie Ostropaceae", *Folia Geobot, phytotax. Praha* 12: 313-16.
Vezda, A. and Wirth, V. (1976) "Zur Taxonomie der Flechtengattung *Micarea Fr. em Hedl.", Folia Geobot. phytotax. Praha* 11: 93-102.

Weber, W.A. (1967a) "Environmental modification in crustose lichens II. Fruticose growth forms in *Aspicilia*", *Aquilo ser. Bot.* 6: 43-51.
Weber, W.A. (1967b) "A synopsis of the North American species of *Cyphelium*", *Bryologist* 70: 197-203.
Weber, W.A. (1968) "A taxonomic revision of *Acarospora* subgenus *Xanthothallia*", *Lichenologist* 4: 16-33.
Weber, W.A. (1977) "*Placynthium* (Ach.) S.F. Gray, a genus of lichens previously unreported from Australia", *Muelleria* 3: 250.
Weber, W.A. and Wetmore, C.M. (1972) "Catalogue of the lichens of Australia exclusive of Tasmania", *Beih. Nova Hedwigia* 41: 1-137.
Wetmore, C.M. (1960) "The lichen genus *Nephroma* in North and Middle America", *Publs. Mich. St. Univ. Mus., ser Biol.* 1(11): 369-452.
Wetmore, C.M. (1963) "Catalogue of the lichens of Tasmania", *Rev. Bryol. et Lichenol.* 32:223-64.
Wetmore, C.M. (1970) "The lichen family Heppiaceae in America", *Ann. Mo. bot. Gdn.* 57: 158-209.
Wiggers, F.H. (1780) *Primitiae Florae Holsatiae* (Bartsch, Kiliae).
Willdenow, C.L. (1787(*Florae Berolinensis Prodromus secundum systema Linneanum ab illustr. C.P. Thunbergio emendatum conscriptus* (W. Vieweg, Berlin).
Wirth, M. and Hale, M.E. (1963) "The lichen family *Graphidaceae* in Mexico", *Contr. U.S. natnl. Herb.* 36(3): 63-119.
Wirth, M. and Hale, M.E. (1978) "Morden-Smithsonian expedition to Dominica: The Lichens (Graphidaceae)", *Smithsonian contr. Bot.* 40: 1-64.
Yoshimura, I. (1971) "The genus *Lobaria* of Eastern Asia", *J. Hattori Bot. Lab.* 34: 231-364.
Yoshimura, I. (1973) "Notes on *Gymnoderma melacarpum* comb. nov.", *J. Jap. Bot.* 48: 283-88.
Yoshimura, I. and Sharp, A.J. (1968) "A revision of the genus *Gymnoderma*", *Am. J. Bot.* 55: 635-40.
Zahlbruckner, A. (1903-7) "Lichenes (Flechten). B. Specialler Tiel", in *Die Naturlichen Pflanzenfamilien* (A. Engler and K. Prantl, Eds) 1: 49-249.
Zahlbruckner, A. (1941) "Lichenes Novae Zelandiae", *Denkshr. Akad. Wiss. Wien* 104(5): 1-132.
Zopf, W. (1903) "Vergleichende Untersuchungen uber Flechten in Bezug auf ihre Stoffwechselprodukte. Erste Abhandlung", *Bei. Bot. Central* 14: 95-126.
Zschacke, H. (1933a) "Familie Verrucariaceae", *Rabenh. Krypt.-Fl.* 9: 1(1), 46-589.
Zschacke, H. (1933b) "Familie Dermatocarpaceae", *Rabenh. Krypt.-Fl.* 9, 1(1): 590-645.

Glossary

acicular: long and needle shaped.
adnate: closely attached, the thallus close to the substrate, or the lower surface of the apothecium fused to the thallus.
aggregated: grouped together.
anastomosing: with numerous cross-connections (Fig. 3A).
anticlinal: at right angles to the surface.
apical: situated at the tip.
apothecium (apothecia): disk shaped fruiting body (ascocarp) of an ascomycetous fungus, whether lichenized or not (Figs 1 and 4).
appressed: closely pressed to the substrate.
arachnoid: cobweb-like in structure.
areole: a small part of a lichen thallus separated from the rest of the thallus by minute cracks.
areolate: consisting of a mosaic of areoles.
articulate: broken into jointed portions and so appearing rather like a crustacean's leg.
ascending: rising from the substrate.
ascus (asci): sac containing spores (Fig. 2A to C).
axis (axes): the main stem of a branched lichen, or in *Usnea*, the central cartilagenous core.
bacilliform: like a short rod.
basidia: inflated hyphal segments bearing basidiospores.
bicornute: curved and drawn into two points.
bilateral: concerning both sides.
bitunicate: ascal wall layers separating to show separate endoascus and exoascus. The fissitunicate ascus of some authors.
blastidia: small subsidiary locules in a thick-walled spore.
bulbate: with a bulbous swelling.
carbonaceous: black, opaque.
cephalodia: thallus structures containing a second phycobiont, usually a blue-green alga.
chondroid axis: the elastic, cartilagenous central core in the genus *Usnea.*
cilium (cilia): a hair-like growth on the margin or upper surface of lobes or on the margin of the apothecium.
clavate: club-shaped.
coccoid: organized in small, more or less spherical groups.
corona: crown-like radiating structure.
cortex: the outermost layer of the thallus which, if present, consists of compacted hyphae which may appear either fibrous or cellular.
crustose: thallus type forming a strongly adherent crust over the substrate: without a lower cortex or rhizines (Fig. 1C).
cupular: like a cup.
cymose: in a sequence one above the other, youngest at the top.
cyphella (cyphellae): a pore in the lower surface, lined with a pseudocortex.
dichotomous: branched into two equal portions, sometimes repeatedly.
ecorticate: without a cortex.
effigurate: without a defined form, grossly irregular (for example not ovoid).
endoascus: the inner layer of a bitunicate ascus.
endobasidial: a fulcrum type normally with short cells and producing pycnospores terminally. The endobasidial fulcrum usually produces bacilliform pycnospores (Fig. 3J and K).
endolithic: within rocks.
endophloic: within bark.
epiphloic: on bark.

epithecium: upper layer of the hymenium, above the asci (Fig. 1A and B).
erect: rising vertically from the substrate.
exciple: the layer surrounding the hymenium which sometimes develops into a distinct margin. A proper exciple has no algal component and is derived from apothecial tissue. A thalloid exciple has an algal component and is derived from the vegetative thallus (Fig. 1A and B).
exobasidial: a fulcrum type normally with long cells producing lateral pycnospores. The exobasidial fulcrum often produces markedly elongate or filiform pycnospores (Fig. 3H and I).
expanded: broadened or extended.
fenestrate: with small perforations.
filamentous: hair- or thread-like.
foliicolous: growing on leaves.
foliose: a thallus type usually with upper and lower cortices, dorsiventral, flat and somewhat leaf-like (Fig. 1E).
fruticose: a thallus form which is usually erect and rather bushy and shrub-like, commonly radially symmetrical but sometimes dorsiventral (Fig. 1F).
fulcrum: hypha giving rise to pycnospores (Fig. 3H to K).
fusiform: spindle-shaped.
gelatinous: like a jelly, slimy.
goniocysts: small aggregations of algae.
gyrose: in a more or less circular or spiral pattern.
heteromerous: a thallus form in which more or less distinct tissues are present.
homiomerous: a thallus form in which the algal and fungal components are uniformly dispersed.
hymenium: that part of the ascocarp composed of asci and paraphyses in a close arrangement (Fig. 1A and B).
hypha (hyphae): a filament of fungal cells.
hypothallus: a growth of undifferentiated mycelia, sometimes present as a distinct layer below or around a thallus.
hypothecium: the tissue in the apothecium immediately below the hymenium (Fig. 1A and B).
immersed: sunken into the thallus.
inflated: expanded, blown up, often hollow.
isidium (isidia): an outgrowth from the thallus with a cortex which serves as a vegetative dispersal unit.
labia: lateral lip-like structures.
lacerate: irregularly incised to form elongate strips, for instance finely lobed.
laminal: on the surface.
lax: loose, loosely woven.
lenticular: shaped like a lens (Fig. 2F).
leprose: a loose powdery mass of algal cells and fungal hyphae.
lirelliform: having a shape rather like the characters of Chinese writing.
mazaedium (mazaedia): a fruiting body in which ascus walls break down to leave an amorphous spore mass.
marginal: on the edge or margin rather than on the upper or lower surface.
medulla: an internal layer of loosely woven hyphae, between the algal layer and lower cortex.
moniliform: regularly constricted, like a string of beads.
muriform: with transverse and longitudinal walls (of spores) (Fig. 2G).
mycelial: made up of intertwined strands of hyphae.
obovoid: generally ovoid, but broader to the distal end.
ostiole: the narrow opening of a perithecium (Fig. 4I).
paraphysis (paraphyses): a specialized hyphal form, either simple or branched, in the hymenium. The term is here used in a broad sense to cover many structures similar to true paraphyses (Fig. 3A to G).
paraplectenchyma: a fungal tissue with a cellular structure superficially like parenchyma of vascular plants.
peltate: like a shield on a central stalk.
pendulous: hanging.
perforate: with holes through the thallus or into a central cavity.
periclinal: parallel to the surface.
perithecium (perithecia): a flask-shaped fruiting body immersed in the thallus with a single, terminal opening (Fig. 4I to K).
phycobiont: the algal component of a lichen.
plaque: like a small disk or plate.
plectenchyma: parenchyma-like tissue of tightly packed hyphae.
podetium (podetia): an erect portion of a thallus derived from tissue of apothecial origin, rising from a primary thallus and bearing apothecia.
polarilocular: referring to spores which are not-septate, but in which the wall thickens, almost dividing the cell into two polar compartments (Fig. 2J).

primary thallus: the first formed thallus which may later give rise to podetia (secondary structures).
proper exciple: an exciple formed from tissue of apothecial origin — without algae in the rim (Fig. 1B).
prosoplectenchyma: a fungal tissue with a cellular structure superficially like collenchyma of vascular plants.
prothallus: a purely fungal layer upon which an algal layer may later develop, commonly black, often found beneath crutose lichens.
pseudocyphella (pseudocyphellae): areas of the thallus in which the upper or lower cortex is missing and medullary hyphae come to the surface.
pseudoparenchymatous: having the appearance of parenchyma, that is isodiametric cells.
pseudopodetium (pseudopodetia): a podetium-like structure that has its origin in vegetative rather than reproductive tissue.
pseudothecium: the fruiting structure of an ascolocular ascomycete.
pycnidium (pycnidia): an often flask-shaped structure producing pycnospores.
pycnospores: the spore type produced in a pycnidium: the function of these is not known (Fig. 3H to K).
pyriform: pear-shaped.
regular: uniform, even, repeating.
reticulate: with a network of rides, lines or cracks on the surface (of a surface); like a net, anastomosing (of paraphyses) (Fig. 3A).
rhizines: a short branch of the thallus resembling a root.
rhizoid: hyphal structures on the lower surface anchoring the thallus.
scrobiculate: marked by shallow depressions.
septate: divided by a septum or wall (Fig. 2).
sessile: without a stem or stipe, sitting on the surface.
simple: unbranched (of paraphyses); non-septate (of spores).
soredium (soredia): a group of algal cells and hyphae, without a cortex, which functions as a vegetative reproductive unit.
squamule: a small scale-like thallus segment, with an upper cortex but lacking a lower cortex (Fig. 1D).
squamulose: a growth form composed of squamules.
squarrose: having numerous short laterals.
stellate: star-like.
sterigmata: spines on basidia which bear basidiospores.
stipe: a stalk that supports a fruiting body.
stipitate: elevated on a stipe.
striate: with parallel stripes or lines.
stroma: a compact mass of fungal tissue in which fruits are formed.
stylospores: large spores produced in pycnidia or similar structures.
terete: circular in cross-section.
terminal: ending the structure.
thalloid exciple: an exciple in which part of the vegetative thallus has grown up around the apothecium so that algae occur in the margin (Fig. 1A).
thallus: the lichen body, both alga and fungus.
tholus: a dome-like structure at the apex of many unitunicate asci.
tomentum: a dense woolly covering of hyphae.
trichotomous: branching equally into three, sometimes repeatedly.
truncate: cut short at the end.
umbilicate: with a point attachment at the centre of a thallus.
unitunicate: ascal wall not developing a distinct and separate endoascus and exoascus: the non-fissitunicate ascus of some authors.
urcelolate: shaped like an urn.
vermiform: like a worm, sinuous.

Taxonomic Index

(Algal genera are differentiated from fungal genera by an asterisk)

Chemical Index